CARPENTRY
METRIC EDITION

BUILDING CRAFT SERIES

BRICKWORK	*By* W. B. M^cKAY
CARPENTRY	*By* W. B. M^cKAY
JOINERY	*By* W. B. M^cKAY
BUILDING GEOMETRY	*By* W. J. STONE

BUILDING CONSTRUCTION
By W. B. *and* J. K. M^cKAY

Volumes One, Two, Three and Four
(*These are also available as two separate books in
hard covers*)
Volumes One and Two together
Volumes Three and Four together

BUILDING CRAFT SERIES

CARPENTRY

By

W. B. M^c KAY

M.Sc.Tech., M.I.Struct.E.

Former Registered Architect and Chartered
Structural Engineer, and Head of the
Department of Building and Structural
Engineering in the University of Manchester
Institute of Science and Technology

FIFTH EDITION (IN SI METRIC UNITS)

BY

J. K. M^c KAY

B.A., B.Sc.Tech., A.R.I.B.A., C.Eng.
M.I.Struct.E., F.F.B.

WITH DRAWINGS BY THE AUTHORS

LONGMAN
LONDON and NEW YORK

LONGMAN GROUP LIMITED

London

*Associated companies, branches and representatives
throughout the world*

Published in the United States of America
by Longman Inc., New York

© J. K. McKay 1966, 1969
© Longman Group Limited 1975

First published . . 1945

Second edition . . 1955

Third edition . . 1966

Fourth (metric) edition 1969

Second impression 1971

Fifth (metric) edition . 1975

Second impression 1976

ISBN 0 582 42541 7

*Printed in Hong Kong by
Wing King Tong Co Ltd*

PREFACE TO THE FIFTH (METRIC) EDITION

THE fourth edition of this book, published in 1969, was the first of its kind to appear in metric terms. At that time the Building Regulations and other legislation had not been metricated and the author was obliged to make assumptions which have since proved to be of reasonable accuracy.

In this new edition the few anomalies resulting from this forecasting have been removed so that the technical content is up to date and in line with official documentation which is well established now in metric format.

Revisions of various kinds have been made and a new section included on trussed rafter roofs which have superseded, almost entirely, purlin roofs for domestic work. A new final chapter on adhesives has been added.

1975   J. K. McKAY

PREFACE TO THE FIRST EDITION

THIS book on Carpentry and its companion volume on Joinery cover the syllabuses in theoretical Carpentry and Joinery in the first and second years of the courses provided in Technical Schools and Colleges in preparation for the examinations of the City and Guilds of London Institute. It is hoped that the two books will meet the needs of students attending these courses and those of others attending craft courses at the various Government Training Centres.

A chapter on drawing has been included in this book with a view to assisting students to attain a satisfactory standard of draughtsmanship at an early stage, an important part of their training being to prepare drawings to scale and to make neat and well proportioned sketches.

As the subjects of conversion and seasoning are dealt with in this volume, a chapter has been given to a description of the manufacture, characteristics and uses of plywood and laminboard, although laminated materials are chiefly employed in joinery. Applications will be found in the book on Joinery which is included in this Building Craft Series. Laminated materials are usually deferred to a more advanced stage, but it has been thought desirable to introduce them in these two books, especially as their use in building is likely to increase after the war.

The author has had the assistance of several experienced craftsmen and teachers of the subject, and in particular he is indebted to Mr J. R. Holland and Mr E. Spencer for revising proof sheets and making several valuable suggestions.

1945   W. B. McK.

CONTENTS

PAGE

PREFACE TO THE FIRST AND FIFTH EDITIONS   5

INTRODUCTION 9

CHAPTER

I. TIMBER 11

Classification, structure, growth, felling, conversion, seasoning, preservation, defects, characteristics and uses of timbers.

II. PLYWOOD 44

Manufacture, characteristics and uses of plywood, laminboards, blockboards, battenboards, composite boards and wall boards. Plywood box beams.

III. DRAWING EQUIPMENT AND DRAWING 54

Instruments, paper ; hints on draughtsmanship ; methods of representation ; lettering, sketching, inking-in, tracing and colouring.

IV. FLOORS 71

Boarded and joisted floors ; Single floors, sizes and spacing of joists, ventilation, trimming, strutting, boarding, joints ; double and triple floors ; double-boarding.

V. ROOFS 112

Terms. Single roofs, including flat, lean-to, double lean-to, couple, close couple and collar roofs. Double roofs ; dormer window ; hips and valleys. Trussed rafter roofs and built-up roof trusses ; glued laminated portal frame. Hyperbolic paraboloid roof. Roof trimming. Flush, open, closed and sprocketed eaves. Determination of bevels. Erection. Gutters.

VI. PARTITIONS 166

Details of stoothed and framed partitions.

VII. SOUNDPROOFING 171

Insulating materials and their application to floors and partitions.

VIII. TEMPORARY CARPENTRY 177

Timbering to trenches. Centering for arches up to 3 mm span. Raking, flying and vertical shoring. Formwork for concrete structures.

CONTENTS

CHAPTER   PAGE

IX. NAILS, SCREWS AND BOLTS 197

    Wire and wrought nails, brads, panel pins, needle points, screws,
    corrugated saw edge fasteners, bolts, timber connectors and dogs.

X. TOOLS 202

    Marking and setting out tools, cutting and planing tools, boring
    tools, impelling tools, abrading tools, cramping and holding
    appliances, and miscellaneous tools and equipment. Portable
    power tools. Care of tools.

XI. ADHESIVES 225

INDEX 229

INTRODUCTION

THE building craftsman in wood is called a carpenter or joiner, the latter name being more often used nowadays. Formerly, the carpenter was occupied in making vehicles, such as carts and carriages, furniture and wood household utensils, in addition to being employed in the construction of buildings. The joiner was at first engaged on upholstering furniture ; gradually his work included the fitting together (or " joining ") of wood members of furniture ; a further development affecting the occupation of the joiner was the making of doors, windows, etc., and the fixing of these in buildings.

There are several broad differences between carpentry (the craft of the carpenter) and joinery (the craft of the joiner).

Carpentry includes permanent structures such as floors (see pp. 71-111), roofs (pp. 112-165), partitions (pp. 166-170) and lintels (wood members at heads of door and window openings). It also embraces temporary work including timbering for trenches or similar excavations (pp. 177-180), centering for arches (pp. 180-185), shoring (pp. 185-193) and formwork (pp. 193-196). Fencing, both temporary and permanent, is also classed as carpentry. Much of this heavier class of constructional work is required to support loads, comparatively little labour is expended on it as much of the timber is left rough from the saw, and a good deal of this labour is done on the building site.

As indicated above, joinery includes the lighter forms of woodwork, such as doors, windows, staircases, cupboards, and panelling for covering walls. Its primary object is to increase the habitability and appearance of a building, most of the work is framed together in the workshop ready for fixing in the building, and as the timber is usually dressed or planed the labour expended on it is large compared with its volume.

Although the craftsman in wood is more often referred to as a " joiner " than a " carpenter ", he should be fully conversant with both carpentry and joinery, as he is now generally called upon to do both.

Timber was formerly sawn, dressed and moulded by hand. Whilst most of this labour is now done by machinery, it is nevertheless essential that the craftsman be well qualified in the use of hand tools, as he is often required to do work which necessitates their use. He should also be able to use the portable power tools (pp. 220-223) which are useful on the building site where machinery is not available. In the larger workshops equipped with a big variety of woodworking machines, those who operate them are called woodworking machinists. These specialists work the machines but are not required to fix in buildings or on sites the wood members which are machined. In the smaller workshops, where only one or two machines may be employed, the carpenter or joiner is often called upon to operate the machines. A knowledge of woodworking machinery is therefore an additional qualification of the craftsman. Some woodworking machines are described on pp. 19-21 ; others are described and some are illustrated in the companion book " Joinery " by the same author.

The qualified craftsman should (1) be able to identify a wide range of timbers (or at least those in common use) and know the relative strengths and other characteristics in order that he may select those most suitable for specific purposes ; (2) be acquainted with the building regulations to ensure that wood members of adequate size are used for floors, roofs, etc., and be able to determine sizes when not specified ; (3) be capable of deciding upon the best joints to adopt having regard to the loads to be supported and the shrinkage and expansion of adjacent members ; (4) have a knowledge of the processes of seasoning and preservation ; and (5) be able to distinguish the several defects to which timber is subjected. See Chapter One.

It is also desirable that he should have a sound knowledge of the work done by craftsmen in other trades, such as brickwork, masonry, slating and plumbing. During the erection of a building he is frequently required to work in conjunction with other trades, i.e., in connection with the bedding of wall plates (p. 75), trimming roofs at chimney stacks (pp. 134 to 137), notching joists for pipes, etc. He should certainly be aware of general defects or omissions which may affect adversely his own work, such as damp walls, the omission of horizontal damp proof courses (p. 79), or the absence of ventilating bricks (p. 80) or site concrete (p. 80), any or all of which may result in ground floor, etc., timbers becoming affected with dry rot (pp. 31-33).

He should be able to read plans, besides having a knowledge of the building regulations. The ability to prepare progress, etc., reports and sketches of constructional details is an additional qualification.

UNLESS NOTED OTHERWISE ALL DIMENSIONS IN THE FIGURES ARE GIVEN IN MILLIMETRES.

CARPENTRY

CHAPTER ONE

TIMBER

Classification, structure, growth, felling, conversion, seasoning, preservation, defects, characteristics and uses of timbers.

CLASSIFICATION

TIMBERS used for building purposes are divided into (*a*) softwoods and (*b*) hardwoods. This division has been established by long usage and is not in accordance with the relative hardness of the timbers, as some softwoods are harder than certain hardwoods.

(*a*) The softwoods are of the *conifer* class. Conifers, with few exceptions, are evergreen trees, *i.e.*, they have leaves throughout the year; the leaves are needle-like. They bear conical sheathed seeds called cones. Most of the timber used for carpentry is of this class, as it is sufficiently strong for most purposes, is often easily worked (see pp. 38-40) on account of its softness and straightness of grain (p. 35) and is relatively cheap. Pines, firs and spruces are examples of conifers. The characteristics and uses of some softwoods are described on pp. 38-40.

British Standard Code of Practice, 112 " The Structural Use of Timber in Buildings " classifies structural softwoods into three groups : Group S1—Douglas Fir, Long leaf Pitch Pine, Short leaf Pitch Pine and European Larch ; Group S2—Canadian Spruce, Redwood, Whitewood, Western Hemlock and Scots Pine ; Group S3—European and Sitka Spruce, Western Red Cedar. Each group is divided into a basic grade of the highest strength and four grades of lower strengths according to the number of growth rings (Fig. 1) per 25 mm ; the greater the number of rings per 25 mm the stronger is the timber. Thus a typical specification for a type of timber which is widely used would read " S2 timber, grade 65 " which denotes any timber in group S2 of grade 65 (i.e. a minimum of 6 growth rings per 25 mm) and CP 112 specifies values of the permissible stresses to be used in the structural design of the timber members.

There are two methods given in the Code of determining the strength or stress grade of a piece of timber : (1) by visual inspection and (2) by mechanical means. Method (1) sets out an elaborate system of visual inspection to analyse the defects in timber and so assess the quality and stress capability; it provides for a General Structural grade (G S) and a Special Structural grade (S S). It is a costly and not very satisfactory method of grading and is being replaced by method (2).

The latter comprises mechanical stress grading where the piece of timber passes through a machine which provides a small load to be applied to it over successive short spans of 1·8 m throughout its length. The resulting deflections are monitored by computer and from them the stiffness and therefore the strength is deduced at intervals. At each measurement the piece receives a short spray of colour which differs according to the grade achieved. As the end of the piece leaves the machine it is dyed with a longer band of colour of the weakest grade noted, or stamped with the appropriate BS grade ; the piece is then said to be of this stress grade. The two grades provided are Machine General Structural grade (MGS) and Machine Special Structural grade (MSS)

(b) The hardwoods belong to the broad-leaf class. Most of them shed their leaves in the autumn and are called *deciduous*. They include the oaks, birches and mahoganies (see pp. 40-42). Some hardwoods, such as oak and teak, are used for carpentry because of their high strength and durable (lasting) qualities, but most of them are employed for joinery on account of their good appearance, hardness, etc. See pp. 40-43.

STRUCTURE

The structure (arrangement of the various parts) of a tree is of a complex cellular character. A portion of a tree is shown in diagrammatic form in Fig. 1. The chief structural parts are indicated at the cross radial and tangential sections.

This shows (1) a central core of fibrous (thread-like) woody tissue (woven particles) called the *pith* or *medulla* which disappears in time, (2) comparatively dark coloured inner concentric rings of woody tissue called *heartwood* or *duramen*, (3) outer and lighter coloured rings of woody tissue called *sapwood* or *alburnum*, (4) radial narrow bands of tissue called *medullary rays* (abbreviated to *rays*) or *transverse septa* (partitions) which radiate from the centre, and (5) the *bark*.

The irregular concentric rings of tissue, forming the heartwood and sapwood, are called *annual rings* (as normally in the temperate climate of this country one ring is formed annually) or *growth rings*. The rings vary in thickness ; thus, a narrow ring produced during a droughty season may adjoin a relatively wide growth ring formed under more favourable climatic conditions ; irregularity in the thickness of a ring will be caused if the tree is exposed to more sun on one side than the other. A ring is divided into an inner portion called the *spring wood* (as it is produced during the spring) and an outer and usually darker portion known as the *summer wood*. The cells of the spring wood have relatively thin walls and large cavities, whilst those of the summer wood have thicker walls and smaller cavities; hence the cause of the sharp contrast between the light coloured spring wood and darker coloured compact summer wood which is clearly defined in many timbers. Whilst some timbers have annual rings which are very distinct, others have rings which are indistinct and there is no contrast between the spring wood and summer wood. In most timbers the medullary rays are hardly visible to the naked eye and are only perceptible through the microscope, although in certain woods, such as oak, they are very conspicuous.

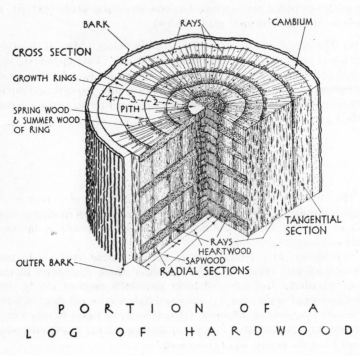

BARK · RAYS · CAMBIUM
CROSS SECTION
GROWTH RINGS
SPRING WOOD & SUMMER WOOD OF RING
4 · 3 · 2 · 1 · PITH
TANGENTIAL SECTION
RAYS
HEARTWOOD
SAPWOOD
OUTER BARK
RADIAL SECTIONS

PORTION OF A
LOG OF HARDWOOD

Fig. 1.

A portion of an annual ring of a softwood is shown in cross-section in Fig. 2 and indicates the honeycombed nature of the structure. This is greatly enlarged, for the number of rings may vary from one to sixteen per cm. It will be seen that the tubular cells, called *tracheids*, are arranged in rows separated at intervals by the medullary rays. The tracheids are approximately 3 mm long and run longitudinally (parallel to the trunk). The sketch shows the comparatively open nature of the spring wood due to the large

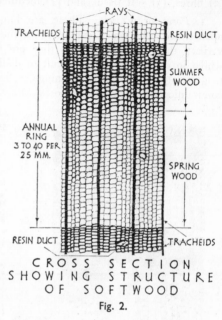

CROSS SECTION
SHOWING STRUCTURE
OF SOFTWOOD

Fig. 2.

RAYS IN SOFTWOODS ARE USUALLY ONE CELL WIDE, SEVERAL CELLS HIGH AS SHOWN & INVISIBLE TO THE NAKED EYE. BROAD RAYS IN HARDWOODS ARE SEVERAL CELLS WIDE; IN OAK THEY MAY EXCEED 20 WIDE & BE SEVERAL HUNDRED (AT LEAST 25 mm) HIGH.

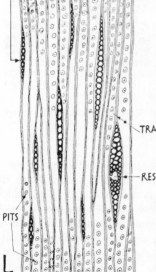

TANGENTIAL SECTION OF
SOFTWOOD SHOWING RAYS

Fig. 3.

cells and thin walls, and the denser summer wood formed by the smaller cells and thicker walls. The cells communicate with each other through holes, called *pits*, in their sides. The medullary rays are straight, narrow (one cell wide and several cells high), radial bands. Fig. 3 shows these rays as they appear in a tangential section; the pits are also indicated. A ray also consists of cellular tissue, called *parenchyma*; the cells are thin-walled and communicate with the adjacent annual ring tracheids. *Resin ducts* (see Figs. 2 and 3) or canals are present in small numbers in the summer wood and rays of certain softwoods (such as pitch pine, redwood and Douglas fir—see pp. 38 and 39); they receive the resin, which is a waste product, secreted by cells immediately surrounding them.

The structure of hardwoods is more

complicated than that of softwoods. It is composed of (1) pores, (2) fibres, (3) soft tissue and (4) medullary rays.

1. *Pores or Vessels.*—These are long vertical tubes composed of open-ended cells which extend down the trunk. The size of the pores varies; thus, in some woods they are conspicuous to the naked eye, whilst in others they are difficult to distinguish even with the aid of a magnifying glass. Small openings, called *pits*, at the sides permit of communication between the pores. Some hardwoods, such as oak, elm

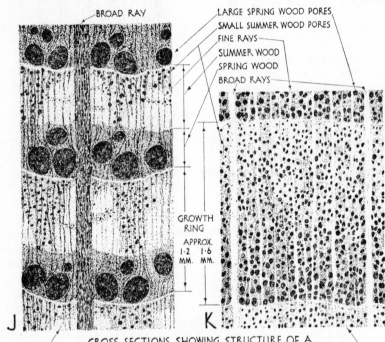

CROSS SECTIONS SHOWING STRUCTURE OF A
RING POROUS HARDWOOD (OAK) & A DIFFUSE POROUS HARDWOOD (BEECH)

Fig. 4.

and ash, have relatively large pores concentrated within the spring wood and smaller pores distributed throughout the summer wood; these are called *ring-porous* woods and an example is shown at J, Fig. 4, which shows a cross-section through a piece of oak. Other hardwoods, such as birch, beech and mahogany, have the pores fairly uniformly diffused (scattered) over the whole growth ring and are said to be *diffuse-porous*; the cross-section at K, Fig. 4, illustrates this, the pores gradually decreasing in size with a maximum in the spring wood. *Softwoods are without pores.*

2. *Fibres* are thick-walled cells and are shorter but similar to the summer wood tracheids of softwoods. The bulk of the wood consists of fibres.

3. *Soft Tissue* or *Parenchyma* consists of thin-walled cells which occur as strands surrounding the pores and as fine lines separating the growth rings.

4. *Medullary Rays* or *Rays*, unlike those of softwoods, are several cells wide and may exceed 2·5 cm in height.

IDENTIFICATION.—The experienced craftsman, by observing such general characteristics as colour, smell, appearance of the rings, rays, etc., has no difficulty in identifying the more commonly used timbers. The use of a hand magnifying glass is useful when a close examination of the structure is necessary. The identification of many timbers cannot be determined with certainty unless a microscope is employed. For this purpose *slides* are prepared consisting of thin (about 0·025 mm thick) slices of the wood which are stained, glued and pressed on pieces of glass ; a slide is placed under the microscope and the latter (which has a high magnification) is adjusted until the detail of the specimen can be distinctly seen when looking through it. The procedure adopted when examining slides under a microscope is beyond the scope of this book. Briefly, as softwoods are without pores (see p. 14), the presence of pores in the specimen is a clear indication that the timber is of the hardwood class. If the slide is that of a hardwood, the grouping of the larger pores (whether ring-porous or diffuse-porous), the distribution of the smaller pores and the character of the soft tissue rays, etc., are observed ; at the same time a larger specimen of the wood is examined and the colour, weight, characteristics of the growth rings, etc., are noted.

FUNCTION.—In softwoods the function of the spring wood cells or tracheids is to conduct water to the leaves, and that of the summer wood tracheids is to strengthen the tree ; the medullary rays serve as storage accommodation for food which passes through the openings (pits) to the adjacent vertical tracheids for distribution. In hardwoods the function of the pores is similar to that of the spring wood tracheids of the softwoods, the fibres strengthen the tree, and food is stored in the soft tissue and rays.

GROWTH

The growing season in this country is from April to September. Moisture, salts, etc., are absorbed from the soil by the roots of the tree, and in the early spring these ascend through the spring wood cells or pores to the branches to develop the leaves which convert the absorbed material, called *sap*, into liquid food for the tree. During the growing period new wood is produced by a thin layer of cells called the *cambium* and situated between the bark and the last formed annual ring (see Fig. 1). These cells divide and subdivide, forming new cells on the inner side which develop into the spring wood of the next annual ring, and new cells on the outer side which gradually grow into the new bark. In the late summer and early autumn the food descends between the spring layer and the bark, to form the denser summer wood of the annual ring. As stated on p. 15, the cells in the rays and soft tissue act as reservoirs for tree food and this is distributed to the tracheids and pores.

FELLING

Timber. for building purposes should be from trees which are felled as soon as possible after reaching maturity. If cut prematurely, the wood is not so durable and contains an excess of sapwood ; if cut long after it has fully developed, the tree produces timber which is brittle and the central portion especially may show evidence of decay. The time taken before trees reach their prime may vary from fifty years (such as ash) to a hundred years (*e.g.*, oak). The best time for felling trees is in the autumn just before the fall of the leaf (when the sap is still thin) or during winter (when the trees contain relatively little sap), as during these periods the evaporation of moisture and the resulting shrinkage (p. 33) are comparatively small.

Trees are felled either by hand or by machine. In this country most tree felling is done by hand. Machines are employed for felling on a large scale ; one type of machine used for this purpose is petrol or electrically driven and resembles that shown in Fig. 8 (see p. 19) with the reciprocating (moving to and fro) saw blade horizontal.

CONVERSION

Immediately after felling, the branches are removed, the trunks are cross-cut into sections called *logs*, and the bark is removed. A log of timber is converted (divided or " broken down ") into various shaped and sized pieces to which the following terms are applied : Baulks, planks, flitches, deals, battens, boards, panels, slices, scantlings, quarterings, strips and fillets. Conversion is done by machines; timber lengths rise in stages of 0·3 m from 1·8 to 6·3 m.

Baulks are square in cross-section of size exceeding 150 mm by 150 mm.

Planks are from 50 to 150 mm thick and at least 300 mm wide.

Flitches are not less than 100 mm by 300 mm.

Deals are pieces of softwood which are from 50 to 100 mm thick by 225 to under 300 mm wide.

Battens are from 50 to 100 mm thick by 125 to 200 mm wide ; *slating* and *tiling* battens are from 16 to 32 mm thick by 25 to 100 mm wide (usually 50 mm by 25 mm or 38 by 19 mm).

Boards are under 50 mm thick by 100 mm and over in width.

Panels and *Slices* are thin wide pieces.

Scantlings are from 50 to 100 mm thick by 50 to 115 mm wide ; this name is often applied to the dimensions of a piece of timber, thus " the joist is of 175 mm by 50 mm scantling."

Quarterings are square sections of from 50 to 150 mm side.

Strips and *Fillets* are under 50 mm thick and less than 100 mm wide.

The conversion of timber should be carried out economically with the minimum of waste. When the converted timber is required to show an attractive figure (p. 36), as for door panels or exposed floor beams (p. 109), the method of conversion adopted is that which will show to the best advantage such characteristics as pronounced annual rings and medullary rays. The several methods of converting a log into planks, boards, etc., are (*a*) rift sawing, (*b*) tangential sawing, (*c*) flat sawing and (*d*) rotary cutting.

(a) *Rift, Quarter or Radial Sawing.*—Rift-sawn timber is defined as that which has been converted so that the annual rings intersect the cut face in *any part* at *more* than 45°. Four forms are shown at A, Fig. 5.

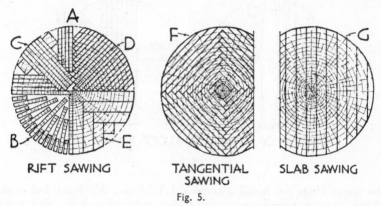

RIFT SAWING   TANGENTIAL   SLAB SAWING
                     SAWING
                  Fig. 5.

That at B is the best if the timber has well-defined medullary rays, as in oak. The log is first sawn into four pieces (or is " quartered ") and each quarter is cut into boards which, like the rays, are radial. The rays appear irregularly on the surface to produce the *silver grain* or *figure* (see p. 37) which is so highly valued for high-class joinery work. It is an expensive form of conversion, as much waste results. More economical methods are shown at C and D, although the latter especially does not show up the figure to the same advantage. Comparatively thicker boards or planks are obtained by the method shown at E. Rift-sawn timber shrinks less in width than flat-sawn (see below) timber and has less tendency to warp (p. 34) or split.

(b) *Tangential Sawing.*—This is shown at F, Fig. 5. It is adopted when the timbers have distinct annual rings (as in pitch pine), as the boards, having their faces tangential to the annual rings, show up to advantage when cut in this manner.

(c) *Flat, Plain or Slab Sawing* is shown at G, Fig. 5. Flat-sawn timber is that which has been converted so that the annual rings intersect the cut face *at least half its width* at *less* than 45°. Compare this definition with that of rift-sawn timber. It is converted either by (1) the *horizontal log band mill* (see Fig. 9) which, starting from the top, cuts the log into flitches, boards, etc., by a succession of horizontal cuts, or by (2) the *vertical log band mill* (p. 20) which breaks down the log by vertical cuts. It will be seen that the outer slabs approximate to tangential cuts and the inner pieces are, in effect, rift-sawn. Timber can be converted quickly, cheaply and with the minimum waste by this method of conversion.

(d) *Rotary Cutting*.—This differs considerably from the above methods and is described in Chapter Two (p. 45).

As stated on p. 33, timber shrinks as its moisture evaporates, and the heartwood shrinks less than the sapwood. The distortion which occurs is shown in Fig. 6. The maximum shrinkage occurs in the direction of

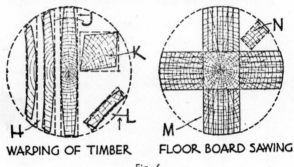

WARPING OF TIMBER   FLOOR BOARD SAWING

Fig. 6.

the annual rings and is called *tangential shrinkage*; it is much less in the radial direction (parallel to the medullary rays) when it is called *radial shrinkage*; it is almost negligible in the direction of its length and is known as *longitudinal shrinkage*. Thus, the thickness of the plank J varies from a maximum at the centre (where there is relatively little moisture in the heartwood) to a minimum at the circumference on account of the larger amount of moisture in the sapwood and the shrinkage which takes place in the direction of the arrows. The piece of quartering, indicated by broken lines at K, is distorted as shown because of the shrinkage in the direction of the annual rings being more extensive than that radially.

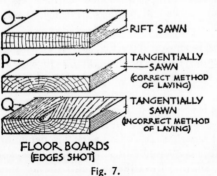

FLOOR BOARDS
(EDGES SHOT)

Fig. 7.

Similarly the plank at L, which is cut tangentially to the rings, shows the shrinkage and distortion (warping—p. 34) which take place. *Tangential shrinkage is approximately double radial shrinkage.* The broken lines indicate the shape of each piece of timber before seasoning.

Thin boards which are used as floor boards (p. 81) should be rift-sawn (as indicated at O, Fig. 7) to give the best results.

This is, however, relatively expensive and the cheaper method of conversion shown at M, Fig. 6 is often adopted ; the remaining portions, consisting of sapwood, are converted into scantlings as required, as at N. Although rift-sawn boards shrink (contract) less and have better wearing qualities, floor boards are often sawn tangentially on the score of economy. Tangentially sawn floor boards should be fixed with the heart side downwards, as at P, Fig. 7 ; if they are fixed with the heart side upwards, there is a tendency for portions to be kicked out, as shown at Q. The floor boards are shown with their " edges shot," *i.e.*, square edges (see p. 82).

WOODWORKING MACHINES

The machines used for converting timber are power driven. Electricity is the chief motive power, although steam, gas and internal combustion engines are used for this purpose. The power may be transmitted from its source either by shafting or, preferably, by a separate motor attached to each machine or group of machines. The shafting (steel rods) is suspended by hangers fixed to the roof or ceiling of the machine shop. Rotary motion is imparted to the shafting by belting which passes over pulleys connected to the shafting and the engine or motor, and the motion of the shafting is transmitted by a belt to the machine. Shafting is gradually being dispensed with as machines with individual motors are installed ; such are known as *motorized machines.*

Logs must be cut into convenient lengths for handling. One of several machines used for this purpose is the *reciprocating cross-cut saw* shown in Fig. 8. The mechanically operated 300 cm long blade has a lower cutting

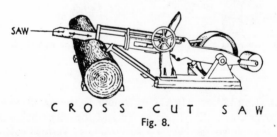

C R O S S - C U T S A W
Fig. 8.

edge, like the ordinary hand cross-cut saw (see 1, Fig. 152), and during its reciprocating motion it cuts downwards with the inward stroke. Logs of large diameter can be rapidly cross-cut by this machine.

There are several types of wood-cutting machines used for converting logs after they have been cut into suitable lengths. These include the (1) horizontal log band mill, (2) vertical log band mill, (3) circular saw mill and (4) band re-sawing machine. The following is a brief description of these power-driven machines :—

1. HORIZONTAL LOG BAND MILL.—This is shown in Fig. 9 and consists of a 150 to 250 mm wide steel band or continuous saw, having teeth

on one edge, which moves horizontally as indicated by the arrows, and is maintained in tension (stretched) over two large (140 to 210 cm diameter) pulleys. The log is supported on a travelling carriage (running on wheels fixed to the floor, or the carriage may be provided with wheels which run on rails) and is fed end-on in the direction of arrow " 1 ". The continuous cutting action of the saw is capable of rapidly breaking down a log or baulk into deals, flitches, boards, slices, etc., by a succession of horizontal cuts, starting from the top. The cross-rail supporting the pulleys is lowered as required after each cut by manipulation of the handwheel. The rate of feed can be varied from 3 to 24 m per minute, and the rate of return may reach 60 m per min.

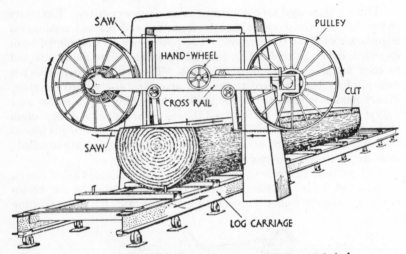

HORIZONTAL LOG BAND MILL

Fig. 9.

2. VERTICAL LOG BAND MILL.—As implied, the band saw has a vertical travel over two pulleys, one above and the other below the log, as it rapidly breaks down the log by a succession of vertical cuts. A log carriage is provided.

Both the horizontal and vertical log band mills are extensively employed when high outputs are required. They are replacing other machines formerly used for converting logs into deals and boards because of their accuracy, high speed and the minimum of waste resulting to the converted timber.

3. CIRCULAR SAW MILL OR RACK FEED SAW BENCH (see Fig. 10).— This consists of a vertical circular saw of 120 to 210 cm diameter and a travelling table (running on rollers) driven to feed the log end-on against the rotating saw as it forms a vertical cut. It is used for breaking down

different sized logs, edging flitches, etc. The rate of feed varies from 3 to 12 m per min, and the accelerated return of the table is 36 m per min.

CIRCULAR SAW MILL
Fig. 10.

4. BAND RE-SAWING MACHINE.—This is similar to but less powerful than the vertical log band mill, and is designed for the rapid (up to 76 m per min) cutting (known as *re-sawing*) of deals, flitches and battens into boards and panels, with the minimum wastage of wood (in the form of sawdust) owing to the thinness of the band saw.

Other machines used for the preparation of timber, and details of the various saws and cutters, are described and illustrated in Chapter One of " Joinery," a companion volume in this Building Craft Series.

SEASONING

Timber cannot be used for either carpenters' or joiners' work immediately it has been felled because of the large amount of moisture (sap) which it contains. This moisture is present as free water in the cell cavities and absorbed water in the cell walls. It is known as the *moisture content* or " *m.c.*" As stated on p. 26, the m.c. is calculated as a percentage of the dry weight of the wood. The m.c. of samples of timber freshly cut from the log may vary from 50 per cent. or more for hardwoods to over 100 per cent. for softwoods. Much, *but not all*, of this moisture must be removed, otherwise the timber will shrink excessively, causing defects in the work and a tendency to decay. The process of reducing the m.c. is called *seasoning* or *conditioning* or *maturing*. Reduction in the m.c. increases the strength, durability and resilience (springiness) of the timber, the wood is lighter in weight, it is easier to work with the saw and other tools, it tends to maintain its size and it is not so liable to split, twist or warp (p. 34).

The amount of moisture remaining in timber *after* seasoning is variable, and depends upon the use to which the wood is to be put. The following are the recommended moisture contents of timber required for various purposes : Good class carpentry work, 20 per cent. (maximum) and rough carpentry work, 25 per cent. ; interior joinery work (including wood block flooring, staircases and wall-panelling which are described in " Joinery "), 9 to 14 per cent. and external joinery work (as for doors and windows described in "Joinery"), 15 per cent. As regards internal joinery work, the aim should be to season the timber until its m.c. is as near as possible to the mean humidity of the building (or room) in which it is to be fixed. Otherwise, if the m.c. is relatively high and the air of the room is warm and dry, a certain amount of moisture will be evaporated and the timber will shrink. Conversely, if comparatively dry timber is exposed to a damp atmosphere it will absorb moisture and will swell. The extent of shrinkage movement in timber may vary from about 8 mm to nearly 16 mm per 100 mm of original width if the m.c. is reduced from 14 to 9 per cent.

METHODS OF SEASONING.—There are three methods of eliminating excess moisture from timber, *i.e.* (1) natural seasoning, (2) artificial seasoning and (3) a combination of natural and artificial seasoning.

1. *Natural Seasoning or Air Seasoning.*—As stated (p. 16), the branches are removed immediately after the trees have been felled, the trees are cut into logs and the bark is removed. If of softwood, the logs are converted (machine-sawn) into baulks and stacked or piled to allow the air to circulate round them. The width of a pile varies from 180 to 360 cm, the height may be up to 500 cm and the length depends upon that of the timber. The timber may be stacked out of doors, when the piles are roughly protected by temporary sloping roofs of a double layer of boards which overhang the sides of the piles. Alternatively, it is stacked in an open shed having a roof and one or more walls. There are several methods of piling, one being shown at c, Fig. 11. A better arrangement is to reduce the number of baulks in every alternate layer; this ensures a better flow of air. Sometimes the piles are arranged with the baulks inclined in the transverse direction; this allows any water which may enter the piles to drain off.

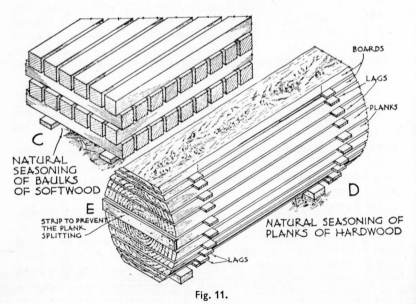

BOARDS

LAGS

PLANKS

C

NATURAL
SEASONING
OF BAULKS
OF SOFTWOOD

D

STRIP TO PREVENT
THE PLANK
SPLITTING

E

NATURAL SEASONING OF
PLANKS OF HARDWOOD

LAGS

Fig. 11.

If required, softwood logs may be converted into boards and stacked as described below.

Hardwood logs may be converted into planks and stacked as shown at D, Fig. 11 ; 12 to 25 mm thick *cross-lags* or *piling sticks* are placed at 1 m intervals ; this is called *sticking*. The converted logs are stacked one above the other. There is a tendency for the timber to split at the ends (see p. 31) ; this will be reduced by coating the ends with paint or, as shown at E, Fig. 11, a thin strip of wood is nailed to each end of each plank ; painting is preferred. If the hardwood logs are converted into boards, the latter may be piled outside or in the shed as shown at A and B , Fig. 12, the piling sticks being spaced at from 60 to 180 cm apart, depending upon the thickness of the boards. Narrow boards may be stacked in two with about 25 mm space between each pair.

The rate of evaporation in natural seasoning is comparatively slow in this country and is only partly controlled. Under average weather conditions it is not possible to reduce the m.c. of timber much below 20 per cent. although it may be reduced to 12 per cent. during a prolonged period of hot weather. As the average reduction is not sufficient for certain internal joinery (see p. 21) it is necessary to dry the timber still further by keeping it in a heated workshop or store before being used.

The length of the drying period in natural seasoning depends upon the temperature and humidity (moisture) of the atmosphere, efficient stacking, and the thickness and density of the timber. On the average, the m.c. in 25 mm thick softwood boards will be reduced to 20 per cent. within two to three *months*, provided they are stacked in spring, and 50 mm thick pieces will dry to a similar amount within three or four months. Hardwoods take longer to season ; thus, 25 mm thick pieces if piled in the autumn will take about nine months to dry to 20 per cent. m.c. and 50 mm thick hardwoods will take approximately a year to dry to the same amount. Any further reduction in m.c. will, of course, take longer and depends upon the heated store to which the timber is transferred. The time occupied in seasoning is reduced if the logs are submerged (as in a river with the thick or butt ends facing upstream), left for about a fortnight to allow the water during its passage through the cells to eliminate some of the sap, when they are removed, converted and stacked.

The advantages of the process of natural seasoning are : (1) It is relatively cheap for small supplies ; (2) it requires little attention ; and (3) defects due to the process are relatively small. The disadvantages are : (1) The rate of drying is very slow ; (2) the process cannot be rigidly controlled ; (3) even under favourable conditions the m.c. cannot be reduced to that required for certain internal joinery ; (4) large stacks of timber require considerable space ; (5) much capital (the cost of the timber) is unproductive for a lengthy period ; and (6) damage to the timber may be caused by fungi (vegetation), especially if the site of the stacks is not adequately drained and covered with ashes or, preferably, concrete.

2. *Artificial Seasoning or Kiln Seasoning.*—This method is employed on a vast scale, as it ensures rapid drying of the timber to any required m.c. under controlled conditions. The timber is stacked in a kiln, of which there are several types, and air, heated to the desired temperature and containing a certain amount of moisture, is circulated through the piles. The air is heated by being passed over steam pipes. This hot air, which accelerates the evaporation of moisture from the wood, must contain a certain amount of moisture, otherwise splitting and case-hardening (see p. 31) of the timber will result. The necessary humidity of the air is obtained by the admission of steam in the form of a spray, and this must be carefully regulated. Adequate circulation of the air is essential, as stagnant air would take up much moisture from the timber and gradually become incapable of reducing the m.c. sufficiently. Hence the air should be of uniform and sufficient velocity ; fresh air must be admitted, and saturated or exhaust air must be removed as required.

A good type of kiln for general work, known as the *external fan compartment kiln*, is shown in Fig. 12. The timber is piled as shown on bearers placed on the floor, the layers being separated by 25 to 40 mm thick piling sticks spaced at from 30 to 90 cm apart, according to the thickness of the timber ; the width of the piles should not exceed 180 cm, as wider stacks make uniform drying difficult ; the face of a pile over the inlet duct is usually inclined as shown at A, as this assists in distributing the warm air through the pile. Sometimes the timber is piled on trucks ; this is economical, as the piling is done outside the kiln, and therefore little time is wasted between the removal of the trucks of dried timber and the charging of the kiln with trucks of unseasoned stuff. As indicated, the air is heated by steam pipes, humidified by sprays, and circulated in the direction of the arrows by a fan situated outside the kiln. Short-circuiting of the air at the end nearest the fan is prevented by the provision of baffles along the air-inlet duct and the adjustment of the dampers at the openings in the inlet and return ducts.

The length of the drying period varies with the size, characteristics and quality of the timber. Approximately the time required in a forced draught kiln (such as that shown in Fig. 12) to reduce the m.c. of 50 mm thick timber from a maximum (green) to 12 per cent. is from one to two *weeks* for softwoods and three to twelve weeks for hardwoods. This is a considerable reduction on the periods given on p. 23 for natural seasoning.

3. *Combined Natural (or Air) and Artificial (or Kiln) Seasoning.*—It is a common practice to reduce the m.c. of timber to approximately 20 per cent. by natural seasoning before subjecting it to further treatment in a kiln. This reduces the above kiln periods to approximately one-third and results in a substantially increased output.

The m.c. is calculated as a percentage of the dry weight of the wood, *i.e.*,

$$\text{Moisture content per cent.} = \frac{\text{Wet weight} - \text{dry weight}}{\text{Dry weight}} \times 100.$$

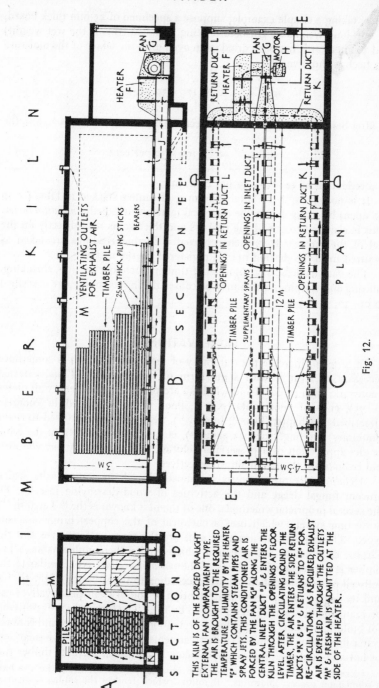

TIMBER KILN

SECTION "D"

M

PILE

K L

SECTION "E E"

HEATER F

FAN G

M VENTILATING OUTLETS FOR EXHAUST AIR

TIMBER PILE

25mm THICK PILING STICKS

BEARERS

J

3 M

PLAN

RETURN DUCT L

HEATER F

FAN G

MOTOR H

RETURN DUCT K

OPENINGS IN RETURN DUCT L

OPENINGS IN INLET DUCT J

SUPPLEMENTARY SPRAYS

TIMBER PILE

OPENINGS IN RETURN DUCT K

TIMBER PILE

12 M

4.5 M

E

D

Fig. 12.

THIS KILN IS OF THE FORCED DRAUGHT EXTERNAL FAN COMPARTMENT TYPE. THE AIR IS BROUGHT TO THE REQUIRED TEMPERATURE & HUMIDITY BY THE HEATER "F" WHICH CONTAINS STEAM PIPES AND SPRAY JETS. THIS CONDITIONED AIR IS FORCED BY THE FAN "G" ALONG THE CENTRAL INLET DUCT "J" & ENTERS THE KILN THROUGH THE OPENINGS AT FLOOR LEVEL AFTER CIRCULATING ROUND THE TIMBER, THE AIR ENTERS THE SIDE RETURN DUCTS "K" & "L" & RETURNS TO "F" FOR RE-CIRCULATION. AS REQUIRED EXHAUST AIR IS EXPELLED THROUGH THE OUTLETS "M" & FRESH AIR IS ADMITTED AT THE SIDE OF THE HEATER.

Thus, taking a simple example, suppose a specimen of 25 mm thick board, 13 mm long weighs 45 gm before being seasoned (this is the wet weight) and 30 gm after it has been dried in an oven until the whole of the moisture has been withdrawn, the

$$\text{m.c.} = \frac{45-30}{30} \times 100 = 50 \text{ per cent.}$$

If, after being partly seasoned, the specimen is re-weighed and is 39 gm, the

$$\text{m.c.} = \frac{39-30}{30} \times 100 = 30 \text{ per cent.}$$

or a reduction of $50-30 = 20$ per cent. m.c.

It is now possible to obtain the m.c. of timber stacked in kilns (or in the open) by using a portable electrical instrument called a *moisture meter*. This is clamped to one of the timbers and the m.c. is read directly on the dial of the meter. It is then removed. Further readings are taken as required until the desired m.c. is registered on the meter.

The various defects or *degrades* of timber, such as checking, shrinking, splitting and warping, which may be caused during seasoning, are described on pp. 30, 31, 33 and 34.

PRESERVATION

In order to increase the durability of seasoned timber it is sometimes necessary to treat it with some form of preservative. Thus, external woodwork, such as doors and windows are painted, and fences, wall plates (p. 75), ends of floor joists (p. 75), floor fillets embedded in concrete (described in Chapter Two, " Joinery "), weather-boarding (used to cover temporary buildings such as garages), etc., are prevented from decaying by the application of a preserving solution. Treatment by *metallic salts* and *creosote* are the two main preservatives.

Metallic salts preservatives are mostly based on copper salts which prevent fungal decay and the activities of wood-destroying insects. Of the several proprietary methods, one of the best known is the " Tanalith " [1] protection system which uses a chemical of the copper/chrome/arsenate type. The copper and arsenical salts are the toxic preservatives and the chrome salts act as fixing agents which render them unleachable. The timber is impregnated in plants of the *vacuum-pressure* type similar to the full-cell process described below. According to the situation of the timber and its degree of exposure, the retention of the dry salt preservatives can be regulated between figures 4 and 20 kg/m^3. The former figure is suitable for building timbers with the higher retentions required for ground contact, water immersion and anti-termite treatments. The less effective treatment of hot-and-cold steeping as given below is also used. Unlike timber that

[1] Executed by Hickson's Timber Impregnation Co. (G.B.) Ltd., who also have a companion process known as " Pyrolith " which renders the timber flameproof (*cf.* p. 29) as well as resistant to decay.

has been creosoted, "Tanalith" treated wood can be painted after it has dried. It is odourless, non-oily and non-staining and it can be used internally where creosoted timber could not be tolerated.

Creosote is an effective general-purpose preservative, particularly for external work. Creosote-treated timber cannot be painted, and so it is usually unsuitable for exposed timber work in internal positions. It is a black or brownish oil and is produced by the distillation of coal-tar, a by-product in gas manufacture. It is poisonous to fungi (one of the principal causes of decay) and insects, permanent when properly applied, cheap and readily available. It is applied by (1) pressure, (2) non-pressure and (3) superficial processes.

1. PRESSURE PROCESSES.—These are generally adopted for treating timber on a large scale and include the (*a*) full-cell process and (*b*) empty-cell process.

(*a*) *Full-cell Process.*—The timber after being seasoned is stacked in a strong steel cylinder, 152 to 300 cm diameter and 460 to 600 cm long, and having a tight-fitting door at each end. The cylinder, fixed horizontally at ground level, has a storage tank containing the creosote (and steam pipes for heating it) connected to it. After the doors have been clamped, the pressure of the air within the cylinder is reduced by means of a pump, hot (varying from 38° to 82° C) creosote is admitted, and a pressure of about 690 kN/m² is applied and maintained for about two hours. The pressure is then released, the creosote is drained off, excess creosote is withdrawn from the timber (by the pump) and the timber is removed. This process is so called because the cells of the wood remain filled with the preservative after the timber has been withdrawn from the cylinder ; it is also known as the *Bethel Process*. The method is best suited for timber which is to be fixed in wet positions (*e.g.*, piles) and where seeping of the creosote (known as " bleeding ") is not objected to.

(*b*) *Empty-cell or Rueping Process.*—The timber in the cylinder is first subjected to a pressure of about 300 kN/m² of compressed air, and this is maintained whilst the hot creosote is admitted and the cylinder completely filled. An additional air-pressure of 70 kN/m² is applied for fifteen minutes or so ; this causes the creosote to enter the wood cells and compresses the air already in them. The pressure is then released to allow the compressed air within the timber to expand and expel much of the creosote from the cells. More creosote is removed from the cells as the pump is applied to reduce the air pressure. Hence the cells are free from the preservative although their walls are well coated with it. This is very suitable for treating timber required for building purposes, as it is effective, the wood is relatively clean to handle on account of the small amount of bleeding which takes place, and it is comparatively cheap, as it only requires approximately half the amount of creosote used in the full-cell process.

2. NON-PRESSURE PROCESSES.—The treatment is known as *steeping*, or *soaking* or *open tank*, and is adopted for relatively small quantities of timber when a pressure plant is not available. The apparatus simply consists of a tank (open at the top, from 366 to 450 cm long, 75 to 150 cm

wide and 90 to 110 cm deep) with steam pipes at the bottom to heat the creosote; in addition, a cylindrical storage reservoir of similar capacity is sometimes provided. There are three methods, *i.e.*, (*a*) hot-and-cold steeping, (*b*) hot steeping and (*c*) cold steeping.

(*a*) *Hot-and-cold Steeping.*—The *seasoned* timber is stacked in the tank and held down by bars at the top. Cold creosote is run in from the reservoir and gradually heated to 94° C; after being maintained at this temperature for one to two hours, the steam is turned off to allow the creosote to cool, the liquid is pumped back to the reservoir and the timber is removed. During the heating period the air in the cells of the timber expands and some of it is expelled; as the creosote cools, the air left in the timber contracts and the partial vacuum created causes the creosote to be gradually absorbed by the timber. The operation only takes a day to complete.

Sometimes the contents of the tanks are re-heated after being cooled and the timber is left to soak in the hot creosote for about two hours before removal. This results in an economy of creosote used, as much of it is expelled from the timber during the re-heating operation.

The pointed ends of fence posts (which are driven into the ground) and the ends of floor joists (which are built into walls—see p. 75) are effectively preserved by *butt treatment*. The timber is placed in a metal drum or barrel with the ends immersed to the required depth in cold creosote which is heated and then cooled as described above. Hot creosote is poured or brushed over the upper portions of the fence posts whilst they are being soaked. The upper ends of the joists are supported by a horizontal piece of timber which is lashed at a convenient height to a vertical post fixed in the ground at each side.

(*b*) *Hot Steeping.*—The timber is soaked in a tank containing the hot creosote for a varying period, depending upon the class of wood, use to which it is to be put, etc. This process is not now often used as it is less efficient, occupies more time and is more costly than hot-and-cold steeping.

(*c*) *Cold Steeping.*—This is even less effective than hot steeping on account of the slow penetration of preservative which takes place, and it is not now recommended.

3. SUPERFICIAL PROCESSES include (*a*) dipping, (*b*) spraying and (*c*) brush application. None is as effective as the aforementioned systems, as the preservative only slightly penetrates the wood. The timber must be seasoned, and the surface should be dry before application. Greater penetration results if the creosote is applied hot.

(*a*) *Dipping* is the best of these surface treatments, except for timber already fixed in position. The pieces of wood are just dipped in a receptacle containing the creosote. The longer the immersion the better.

(*b*) *Spraying.*—The preservative is applied in the form of a fine spray as it is forced through the nozzle of the appliance by compressed air. It is an effective form of surface treatment, as the pressure makes possible the penetration of any cracks or crevices.

(*c*) *Brush Application* is the most common method of treating existing exposed woodwork with creosote or other preservative. The liquid should be applied liberally with the brush, and any cracks in the timber should have special attention. At least two good coats should be given, the first being allowed to dry before the second is brushed on. Where accessible, this treatment should be renewed every three years, especially if it is external work such as gates, fencing and timber outbuildings.

It is very essential that the timber should be properly seasoned before any of the preserving processes are applied, as the presence of moisture impedes the penetration of the preservative. The greater the penetration the more effective the process.

FIRE-RETARDING.[1]—Timber cannot be made fireproof, although there are several chemical solutions and proprietary paints available for rendering it fire-retarding. An effective fire-retardant is ammonium phosphate. The timber should be well seasoned and the material applied by any of the methods described for preservation.

DEFECTS

The defects in timber may be classified according to (a) those developed during its growth, and (b) those occurring after it has been felled. Those which appear during the growth of the tree include deadwood, druxiness, foxiness, coarse grain, twisted grain, cup shakes, heart shakes, knots and upsets. Defects which develop or are produced after the timber has been felled are brashness, checking, collapse, doatiness, dry rot, wet rot, chipped grain, shrinking, swelling, splitting, wane and warp. In addition, considerable damage to timber may be caused by insects such as the deathwatch beetle, the common furniture beetle and the lyctus powder-post beetle.

The following is a brief description of these defects :—

Deadwood.—Applied to redwood (p. 39) which is deficient in strength and weight, and having an abnormal pinkish colour ; is the result of trees being felled after they have reached maturity.

Druxiness is an incipient (early) decay which appears as whitish spots or streaks ; is due to fungi (a form of plant life) which probably gains access through a broken branch and sets up decay.

Foxiness.—Reddish or yellowish brown stains in oak (p. 42) caused by over-maturity or badly ventilated storage during shipment ; is an early sign of decay.

Coarse Grain.—This is applied to timber (see p. 36) which has very wide annual rings caused by the tree growing too rapidly ; wood is generally deficient in strength and not durable.

Twisted Grain or Fibre (see D, Fig. 13).—Fibres are twisted to such an extent that a relatively large number are cut through when a log is converted into planks, boards, etc. ; such converted timber will twist or warp (p. 34) ; caused by the action of the wind on the branches twisting the trunk of the tree.

[1] See also footnote to p. 26.

Cup Shakes or Ring Shakes (see A, Fig. 13).—Cracks or clefts developed between two adjacent annual rings; they interfere with the conversion of timber, resulting in waste; caused by the sap freezing during its ascent in the tree during the spring.

Heart Shakes (see B, Fig. 13).—' begin at the heart of the log; a single cleft is not serious. A *star* consists of several heart shakes somewhat in the form of a star; co n of timber is rendered difficult and uneconomical because of the re g waste. They are an early sign of decay and are caused by shrinka an over-mature tree.

DEFECT: N TIMBER

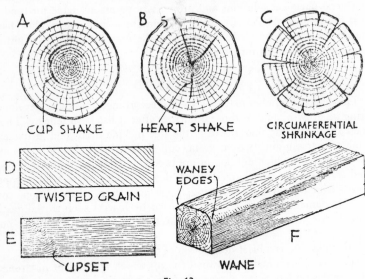

CUP SHAKE   HEART SHAKE   CIRCUMFERENTIAL SHRINKAGE

TWISTED GRAIN

WANEY EDGES

UPSET   WANE

Fig. 13.

Knots are sections of branches present on the surface of wood in the form of hard dark pieces. It is almost impossible to obtain certain converted timbers entirely " free from knots " (as is sometimes specified). Those securely joined to the surrounding wood, and known as " tight knots ", are sound and not objectionable unless large. Wood with " large " or " loose " knots should not be us l s th former especially are unsightly and the latter are readily removed t ntaining many knots is difficult to work. Knots are a source of w ness if present in timber to be used as struts or similar members.

Upsets or Rupture (see E, Fig. 13).—Fibres deformed due to injury by crushing during the growth of the tree.

Brashness or Brittleness may be caused to timber if it has been too rapidly dried or been subjected to a very high temperature in the kiln. Such timber breaks with a short fracture.

Checking is a defect or *degrade* due to seasoning and is the result of unequal drying. The fibres are separated longitudinally, but such separation does not extend throughout the whole cross-section of the wood. The checks are usually much shorter than shakes; the latter are not seasoning defects. As the moisture evaporates more rapidly from the surfaces, they tend to shrink before the inner layers, and splitting results. Checking must not be confused with splitting (see p. 34). The various forms of checks are :—

(a) *End Checks*.—As implied, these occur on the ends; they are caused by the moisture evaporating more quickly through the end grain than other surfaces and the shrinkage being held back by the greater body of wood. Painting the ends minimizes end checking, as this reduces the rate of drying.

(b) *Surface Checks*.—These form on the outer faces during the early stages of seasoning; later they may close and are only exposed on dressing the timber.

(c) *Internal Checks or Honeycomb*.—These appear in the interior of the timber if the drying conditions are too severe in the early stages. The separation of the inner fibres is due to the shrinkage of the dried surface fibres being resisted by the wetter core, and when the internal moisture dries out later the core fibres are prevented from shrinking by the dry outer layers. A condition of stress results, called *case-hardening*, and this produces the short splits or honeycombing.

Collapse is a condition which may occur during the early stages of kiln seasoning very wet timber. The cells in the timber are flattened if the water evaporated from them is not at once replaced by air. Collapse is prevented if the timber is dried at low temperatures in the early stages.

Doatiness or Dote.—Form of incipient decay indicated by patches of greyish stains speckled with black which are relatively soft; due to imperfect seasoning, or badly ventilated storage, and found in American oak (p. 42), beech (p. 40) and birch (p. 40).

Dry Rot.—This disease, which is highly infectious, causes a tremendous amount of destruction in timber. The decay is due to a fungus which feeds upon the wood. Partially seasoned wood fixed in a warm, damp and badly ventilated position is very liable to attack by this fungus. The appearance of affected timber varies with the extent of the disease. In the earliest stage, the spores (germs or seeds) of the fungus throw out minute silky hollow threads which rapidly develop into an interlaced network covering the timber; this covering is grey coloured, relieved with blue or yellow patches. If the conditions are very damp, cotton-wool-like masses, snowy white in colour and with yellow patches, may be formed; these develop into brown or dark red coloured sponge-like growths, called " mushrooms ", which may exceed 30 cm in diameter. Countless numbers of spores are produced on the surface and these can be readily conveyed by air currents, rats, mice and insects to infect timber far removed from the original site; infected tools and clothing may also spread the disease. Greyish white strands, or hollow tubes up to 6 mm thick, may be produced and spread in all directions over timber, brickwork, plaster, etc. ; these convey water to dry timber and thereby provide favourable conditions for the extension of the decay. In this manner dry rot may be transmitted throughout a building. As the fungus develops and feeds upon the cell walls, the timber becomes brittle and falls to powder if pressed by the fingers, it is dry (hence the name applied to the rot), the surface becomes uneven, and cracks extend both with and across the grain to divide it into *cubical* pieces. Sometimes the decay is entirely internal and there is no outside evidence of it.

The following precautions should be taken to prevent the onset of this disease : (1) All timber should be sound, well-seasoned stuff of good quality ; *timber having a moisture content of less than 20 per cent.* (see pp. 21 and 22) *is not likely to be attacked by the disease.* (2) The timber must be kept dry when fixed in buildings ; hence efficient damp proof courses (p. 79) and site concrete (p. 80) must be provided when ground floors are of timber (pp. 71-91), defective roof coverings and gutters or pipes should be at once repaired, wall plates and built-in ends of floor joists should be properly treated with a preservative (p. 28), etc. (3) Adequate circulation of fresh air round all timbers must be provided, as stagnant moist air is particularly favourable to the growth of dry rot ; *through* ventilation under all wood floors (especially ground and basement floors) is therefore essential (see p. 80) ; air spaces round ends of built-in floor joists (p. 80), and the lower ends of roof rafters where walls are beam-filled [1] (Figs. 98 and 100) must be provided. (4) Site concrete (see p. 80) should be well brushed and pieces of wood, shavings, etc. removed before the boarding of ground floors is fixed, as outbreaks have been traced to affected debris of this description which is liable to dampness ; likewise wood setting-out pegs used for concrete foundations and floors should be removed. (5) Linoleum and similar covering should not be laid on new wood floors, especially wood-covered concrete floors (Chapter Two, " Joinery "), before they have had time to dry out.

Dry rot may be detected by the presence of any or all of the following symptoms : (1) The appearance of the fungus described on p. 31 ; (2) " cubical rot " (see above) and other signs of infected timber already referred to ; (3) decay or collapse of wood members (the underside of floor boards and the backs of skirtings may be extensively decayed—the former being readily broken by stamping the heel on them) ; (4) any objectionable musty smell indicating dampness ; and (5) a deposition of red-coloured powder (which teems with the spores) below a floor.

The measures necessary to eradicate the disease depend upon its extent. Drastic steps must be taken in serious cases. Thus, taking a bad case as an example, the following would be necessary if an examination of a ground floor disclosed the decay to be extensive : (1) The whole of the timber (including floor boards, joists, wall plates, and any skirtings with their fixings, such as plugs and grounds) is removed, carefully taken outside and immediately burnt ; any plaster behind which the fungus

---

[1] Beamfilling is brickwork or masonry continued up between and usually to the back of roof spars after the latter have been fixed (see Fig. 98).

may have spread must also be hacked off and removed. (2) Wall faces below the floor, timber pockets and site concrete [1] are well cleaned down with a wire brush and these sweepings (in which the spores are present) are carefully taken and spread over the wood fire and destroyed. (3) A plumber's blowlamp (a hand appliance, which produces a very hot flame, used for making joints in lead pipes, etc.) is applied to the *whole* of the brushed surface to destroy the spores; this should be applied slowly and for a sufficient time until the flame renders the surfaces hot to the touch. (4) An antiseptic, such as a 4 per cent. solution of sodium fluoride or a 5 per cent. solution of magnesium silicofluoride, is liberally applied in two coats to the brushed surfaces. (5) Necessary structural work, such as the insertion of a horizontal damp proof course, the provision of airbricks (p. 75) and holes in division and sleeper walls (p. 80) to ensure *through* ventilation, is carried out. (6) New floor timbers, skirtings, etc., of sound, well-seasoned stuff, are fixed; the joists, underside of floor boards, backs of skirtings, wall plates, etc., are sometimes treated with a preservative before being fixed.

The above operations are costly and are only necessary when the decay is widespread. In a mild case it may be only necessary to remove the decayed timber, together with at least 30 cm of the adjacent sound wood, and replace it with well-seasoned and preserved wood of good quality after brush-treating as much as possible of the existing timber with a preservative.

The species of fungus referred to on pp. 31 and 32 is the *Merulius lacrymans*. Another species which produces dry rot in building timbers is *Coniophora cerebella* or Cellar Fungus. It only attacks wet timber, and the decay is usually confined to cellar, bathroom, etc. floors where there has been a leakage of water, and to leaky roofs. The decay caused by the Cellar Fungus is less serious than that produced by the *Merulius*, as it is at once arrested if the cause of the dampness is attended to and the timber is allowed to dry. The decayed wood is much darker than that affected by the *Merulius*, the cracks are not " cubical " (p. 32) but are mainly with the grain, the strands are brown or almost black, cotton-wool-like masses (p. 31) are absent and mushroom growths are rarely seen.

Wet Rot is a chemical and not a fungoid decay of timber; affected portions are reduced to a greyish-brown powder and these only need to be removed and replaced; caused by timber being subjected to alternating wet and dry conditions.

Chipped Grain is a slight defect caused by a planing machine (Fig. 6, " Joinery ") or tool (p. 208) removing a portion below the surface of the wood as it is being dressed.

Shrinking and Swelling.—When the amount of moisture in timber is reduced during seasoning the wood shrinks (contracts) and, if wood absorbs additional moisture after being seasoned, an increase in volume

[1] If no site concrete exists, the top soil, which may be teeming with spores dropped from the affected timber, is excavated, carefully removed (to prevent droppings(and buried.

(swelling) results. The extent of this movement is referred to on p. 22, and is influenced by the manner of conversion (see p. 18), moisture content and the proportion of heartwood (p. 12). The evaporation of the absorbed water in the cell walls during the process of seasoning does not commence until the whole of the free water (p. 21) in the cavities has disappeared. When the last of the free water has been removed, and the cell walls are still saturated, the condition known as the *fibre saturation point* is reached ; the m.c. at this stage varies from 25 to 30 per cent. Shrinkage does not occur until after the free water has been totally removed and the reduction of the absorbed water commenced ; changes in size and often in shape (p. 18) then occur.

Splitting is the separation of the fibres which extends through a piece of timber from one face to another. Splits are sometimes called *through checks*.

Circumferential Shrinkage (see c, Fig. 13).—Clefts radiate from the circumference of the log towards the centre ; these decrease in width from the outside and are usually limited to the sapwood ; wastage occurs during conversion ; result of shrinkage which occurs during seasoning.

Wane is the original splayed or rounded surface of the tree which remains at the edge or edges of a piece of timber after conversion ; sometimes known as *waney edge*. A baulk with two waney edges is shown at F, Fig. 13. Wane appears at the upper end of the baulk, the lower end being sufficiently large to give a square cross-section. It is due to converting too large a baulk from the log ; not considered detrimental when used in positions where the appearance and large amount of sapwood (p. 12) are unimportant considerations.

Warp or *warping* is the distortion which may occur during shrinkage. The various forms are : (*a*) *Bow* or *bowing* is the slight curvature of a piece of timber, such as a plank or board, in the direction of its length ; (*b*) *cup* or *cupping* is a curvature in the cross-section of a piece of timber ; (*c*) *spring* or *springing* is a curvature of the *edge* of a piece of timber, the face is not affected and is therefore flat ; and (*d*) *twist* or *wind* is a spiral distortion (*winding*) along the length of a piece of timber.

Damage caused by the :—

Death-watch Beetle.—Softwoods and recently seasoned timber are rarely affected, as it chiefly attacks well-matured hardwoods. Very extensive damage has been caused to oak roofs of ancient buildings by the ravages of the death-watch beetle. The beetles lay their eggs in cracks and holes in the wood and the greyish-brown larvæ or grubs which hatch out of them are about 6 mm long when full-grown. The destruction is caused by the grubs boring numerous tunnels about 3 mm diameter in the wood and producing dust in the process.

Dampness and poor ventilation which are associated with dry rot are conducive to attack by the beetle. So treatment of infested timbers must include the provision of adequate ventilation and air spaces round built-in ends of members, and the remedy of defects causing dampness. Badly infected timber must be removed and replaced by sound stuff free from sapwood. Remedial treatment comprises brushing down of affected timber, removal of dust and application of patent insecticide by brush or spray.

Common Furniture Beetle.—These beetles attack both hardwoods and softwoods, and especially unpolished furniture and wall-panelling. The damage caused is somewhat similar to that by the death-watch beetle except that the diameter of the bored holes is only about 1·6 mm. Patent insecticide solutions, which do not discolour tne timber, may be applied as described above.

Lyctus Powder-post Beetle.—Grubs of these beetles cause much damage to furniture and internal joinery such as panelling, and to timber stacked in yards. The sapwood only of certain inadequately seasoned hardwoods (such as ash, elm, oak and walnut—see pp. 40, 42 and 43) is subject to attack ; softwoods and well-matured hardwoods are not affected. White grubs (about 6 mm long), which hatch out of the eggs laid in the pores of the sapwood by the beetle, form small holes (up to about 2 mm in diameter) on the surface of the wood ; the presence of small piles of dust cast out from the holes is also characteristic of infected timber. Such can be treated by several brush applications of one of several proprietary solutions to the surface during the spring and summer (when the beetles developed from the grubs emerge from the timber). Proper seasoning of the timber destroys any grubs which may be present.

House Longhorn Beetle.—This attacks only softwoods ; in this country the occurrence is sporadic but confined mainly to roof timbers in N.W. Surrey in which areas the Building Regulations have provisions requiring timber to be specially treated. It is grey to black in colour varying from 6 to 25 mm long and makes oval holes about 5 mm on the longer dimension.

CHARACTERISTICS AND USES

It is only possible to describe here the characteristics and uses of a few of the many timbers used for building purposes. The softwoods and hardwoods described on pp. 38-43 are some of the more important. Several terms used in this description to express the appearance of the cut surfaces of the timbers include (1) grain, (2) texture and (3) figure.

1. GRAIN applies to the general *direction* of the fibres and cellular units (such as tracheids—p. 13) in relation to the longitudinal edges of a piece of wood. There are several kinds of grain, *i.e.* (*a*) straight, (*b*) irregular, (*c*) wavy, (*d*) spiral, (*e*) interlocking and (*f*) diagonal.

(*a*) *Straight-grained* timber has parallel fibres, is relatively strong, is easy to work (with the tools or machinery) and has a plain figure (p. 36).

(*b*) *Irregular-grained* timber has inclined fibres due to the presence of knots, is relatively weak, is difficult to work, but gives an attractive figure (p. 37).

(*c*) *Wavy* or *curly-grained* timber has fibres which frequently change direction and produce alternating darker and lighter wave-like stripes on the surface (such timber when split has a corrugated surface) ; it is valued because of its highly decorative appearance (p. 37).

(d) *Spiral-grained* timber has fibres arranged spirally ; it is of reduced strength.

(e) *Interlocking-grained* timber has fibres in successive annual rings inclined in opposite directions ; may twist excessively when being seasoned, is not easy to work ; produces an attractive figure (p. 37).

(f) *Diagonal-grained* timber is straight-grained stuff which has been improperly converted so that the fibres are inclined to the longitudinal edges ; this reduces the strength.

Other terms in which grain appears are :—

Flat grain or *slash grain* are applied (incorrectly) to timber which has been flat-sawn (p. 17).

Quarter grain, *edge grain*, *vertical grain* and *comb grain* are loosely applied to quarter-sawn timber (p. 17).

End grain refers to the arrangement of the exposed fibres on the cross-cut surface.

Even grain and *uneven grain* are used to describe timber whose annual rings are either uniform (" even ") or irregular (" uneven ") in width. Such terms are inaccurate (as the appearance is not influenced by the " direction of the fibres "—see definition of grain) and a better description is " annual rings of regular (or irregular) width."

" *Short in the grain* " is applied to timber which fails due to brittleness. This is another inaccurate application of grain, as the condition is not affected by the direction of the fibres. Whilst it may be characteristic of certain timbers, it is also due to improper seasoning and dry rot (p. 31).

2. TEXTURE applies to the size and order or arrangement of the cells ; some of the terms are :—

Coarse textured hardwoods have large pores and/or broad medullary rays, such as oak (p. 42).

Fine textured hardwoods have small pores and/or narrow rays, such as beech (p. 40). Most softwoods are fine textured on account of their small cells.

Medium or *moderately coarse* hardwoods are intermediate grades and include, for example, Honduras mahogany (p. 41).

Even textured timbers include diffuse-porous (p. 14) hardwoods (such as beech) and softwoods with annual rings having slight contrasting spring and summer wood (such as yellow pine—p. 40).

Uneven textured timber include ring-porous (p. 14) hardwoods (such as elm—p. 40) and softwoods having strongly contrasting zones of spring and summer wood (such as Douglas fir—p. 38).

Sometimes softwoods are classified as " coarse grained " (or " coarse textured ") or " fine grained " (or " fine textured ") when their annual rings are wide and narrow respectively ; such expressions are not correct, as the width of the rings does not affect either grain or texture ; they are best referred to as " wide-ringed " or " narrow-ringed " timbers.

3. FIGURE is the pattern on the surface of timber formed by the arrangement of the various tissues, and influenced by the grain and colour. There is a big variation in the quality of the figure ; thus, in general, straight

grained timber has only a plain figure, whereas suitably converted irregular, interlocked or wavy grained timber will show beautiful markings.

The method of conversion affects largely the nature of the figure. When the appearance of the timber is an important consideration, as for wall-panelling, hardwood doors and furniture, the form of conversion adopted should expose the characteristic elements to the best advantage. Thus, when oak is used for such purposes it should be quarter-sawn, in order to disclose on the cut surface relatively large portions of the con-spicuous broad medullary rays which give the richly ornamented figure called " silver grain " ; pitch pine (p. 38) is flat-sawn or, preferably, tangentially-sawn to expose the distinct growth rings of light-coloured spring wood and dark summer wood.

Pronounced irregularities in the direction of the fibres usually give an attractive figure. The following are examples : Irregular grain due to small raised patches on the annual rings produces *blister figure*, which is very decorative and common to pitch pine (p. 38) and sapele (p. 42) ; the irregular grain conforming to small depressions in the annual rings gives the characteristic *bird's-eye figure* of rock maple (p. 42) ; the attractive figure resulting from irregular grain at the *burrs* (swellings at the junctions between the branches and trunk—see A, Fig. 17) in walnut (p. 43) is called *burr figure*. Interlocked grain is responsible for the *stripe figure* or *ribbon figure* peculiar to African mahogany (p. 41) when quarter-sawn and for the *mottled figure* characteristic of *black bean* (p. 40), etc. ; interlocked grain in mahogany, sapele and other tropical timbers is sometimes irregular, and the striped figure is interrupted by longitudinal darker bands to form what is known as *roe figure*. Wavy or curly grain may produce markings in Cuban mahogany (p. 41), rock maple, (p. 42) sycamore (p. 43) and walnut known as *fiddle-back figure* (so called as such decorative wood is usually selected for the backs of violins); this is caused by alternate dense and lighter transverse bands crossing the grain.

Richly decorative figure results when the converted timber is that from just below the *crotch* (fork) or at the *stump* (base) of a tree (see A, Fig. 17) ; thus, the disturbed curly grain produces a figure in mahogany and rock maple known as *feather curl* (as it has the appearance of ostrich feathers). Another modification as a result of contorted grain, sometimes seen in oak, is appropriately named *ram's horn figure*, the short waves forming a series of narrow transverse stripes.

Ripple marks are the transverse sections of medullary rays seen on the cut surface of some hardwoods (including Honduras mahogany and syca-more) and arranged regularly at intervals in straight lines across the width.

Variations in colour affect the figure and increase the decorative value of timber, *i.e.*, the rich colour of walnut streaked with dark bands.

Pith-flecks are discoloured brown streaks or spots occasionally seen in birch (p. 40) and Rhodesian teak (p. 43).

As already stated, the following softwoods and hardwoods are only a selected few of the several hundred timbers used for commercial purposes.

Many of the hardwoods especially are not used for carpentry work, and are included here in order that reference may be made to timbers suitable for joinery. *Home-grown Timbers* are from trees grown in the British Isles, and those obtained from the British Commonwealth are called *Commonwealth Timbers*. The weight of timber is influenced by the m.c. ; the greater the m.c. the greater the weight. The weights stated in the following description are the average when the m.c. is 15 per cent. The timbers are arranged in alphabetical order ; the names given first are so called " Standard Names ", those appearing in brackets are alternatives.

SOFTWOODS

Douglas Fir (British Columbia, Columbian and Oregon pine).—From British Columbia, Western U.S.A. and British Isles. Weighs 530 kg/m³. Pink to light reddish brown ; well defined rings and prominent figure ; straight grained with tendency to wavy or spiral grain ; difficult to work, strong ; available in large sections and long lengths ; stains but does not paint well. Better quality (known as " clear grade ") used for first-class joinery, as for doors, windows, panelling, plywood, floor boarding and blocks ; " merchantable grade " is used for carpentry ; home-grown of inferior quality is used for rough boarding (as for roofs).

European Larch.—From Europe, including British Isles. Weighs 590 kg/m³. Reddish brown heartwood, yellowish white sapwood ; distinct annual rings ; straight grained ; very durable, tough and strong ; resinous ; difficult to work. Most valuable home-grown softwood. Used for carpentry of all kinds ; fencing, gates, wood buildings, scaffolding, etc.

Parana Pine.—From Brazil. Weighs 560 kg/m³. Yellowish buff to darker brown with bright red streaks in heartwood ; indistinct rings ; fairly straight grained ; even textured ; must be carefully seasoned to avoid serious warping, shrinking and splitting ; not durable (and therefore not suitable for external work) and inclined to be brittle (hence unsuitable for floor joists and roofs) ; when properly seasoned is easy to work. Used for internal joinery such as doors, floor boards and plywood cores (see pp. 44-53).

Pitch Pine.—From Southern U.S.A. Weighs 660 kg/m³. Light red ; very distinct annual rings with large proportion of summer wood which gives bold effect, occasionally has blister figure, straight grained ; very strong and durable ; resinous, which affects ease of working ; even textured ; obtainable in long lengths ; subject to heart shakes. Used for good class general carpentry and joinery ; church and school furniture, office fittings, shoring (p. 185).

Red Pine (Canadian, Ottawa and Quebec red pine).—From South-Eastern Canada and Northern U.S.A. Weighs 530 kg/m³. Light red or reddish yellow heartwood, creamy white sapwood ; generally straight grained ; easily worked to a smooth silky finish ; fairly durable. Used for similar purposes as redwood.

Redwood (Scots pine or fir, northern pine, red deal, yellow deal, etc.).—From British Isles, Norway, Sweden, Finland, Poland and Northern Russia. Weighs 530 kg/m³. Pale reddish brown heartwood, light yellowish brown sapwood ; distinct rings ; straight grained ; easily worked to a clean finish ; very durable when preserved ; tough, strong, moderately resinous. • Used extensively for general carpentry and joinery (floors, roofs, doors, windows, etc.). Quality not so reliable as formerly, and Douglas fir (see previous page) is often preferred for first-class work.

Sitka Spruce (silver spruce).—From British Columbia, Western U.S.A. and British Isles. Weighs 470 kg/m³. White to pink ; distinct rings ; mostly straight grained, occasionally spiral grained ; easily worked when free from knots, with satiny finish when planed ; tough ; obtainable in large sizes. Imported used for good-class carpentry and joinery ; home-grown used for temporary work such as formwork for reinforced concrete structures (p. 193). Canadian spruce, from Eastern and Northern Canada, is inferior to silver spruce and the better qualities are used for cheaper joinery.

Sugar Pine.—From California and U.S.A. Weighs 450 kg/m³. Characteristics similar to yellow pine (see below). Used for general internal joinery as a substitute for yellow pine. Siberian pine is also similar to yellow pine, for which it is used as a substitute.

Western Hemlock (grey fir).—From British Columbia and Western U.S.A. Weighs 500 kg/m³. Pale brown ; distinct rings and good figure ; usually straight grained ; fairly even textured ; not durable when subjected to alternate dry and wet conditions. Used for general joinery ; best quality for decorative work (including panelling and furniture), flooring.

Western Red Cedar (Pacific red cedar).—From British Columbia and Western U.S.A. Weighs 390 kg/m³. Reddish brown, weathering to silver grey ; distinct rings ; straight grained ; easy to work ; very durable, brittle. Used for general carpentry and joinery ; roof covering (called *shingles*) and weather-boarding (for covering walls) ; decorative work, including panelling.

Western White Pine (finger cone and mountain pine).—From British Columbia and North-Western U.S.A. Weighs 450 kg/m³. Characteristics similar to yellow pine (see below), but rings are narrower, and it is slightly harder and stronger. Used for purpose similar to yellow pine.

Whitewood (European spruce, white deal or fir, northern and Baltic whitewood).—From Northern and Central Europe and the British Isles. Weighs 430 kg/m³. Yellowish or pinkish white ; distinct rings ; straight grained ; presence of many hard black knots affects working ; smooth silky finish. Used for internal carpentry and cheaper joinery ; rougher grades (including home-grown) for temporary work.

Yellow Pine (Canadian white and yellow, Ottawa white, Quebec, Weymouth and white pine).—From Eastern Canada and U.S.A. Weighs

420 kg/m^3. Pale straw to light reddish brown; rings indistinct; soft, straight grained (with fine short streaks), even textured; easily worked to smooth silky surface; moderately durable. Used for good-class general joinery; first quality expensive and now employed chiefly for pattern making, carving and similar special work. See sugar pine and western white pine.

HARDWOODS

Ash (American and Japanese ash).—From British Isles, Eastern Canada, U.S.A. and Japan. Weighs 700 kg/m^3. White to light brown; ring porous, large pores distinct; growth rings distinct, rays indistinct; straight and coarse grained and occasional decorative burrs; very tough and elastic, not durable when exposed. American and Japanese similar but inferior to home-grown. Figured timber used for decorative work including panelling and furniture; plywood; chiefly used for hammer, etc. shafts, hockey, etc. sticks, motor, etc. body framework.

Beech (American beech).—From British Isles, Central Europe, South-East Canada and North-East U.S.A. Weighs 740 kg/m^3. White and pale brown; diffuse porous, pores barely visible; rings moderately distinct, rays very distinct as flecks; straight grained; fine texture; works easily; hard and very durable if wet or dry. Used for block and parquet flooring, doors, furniture, wood-working tools such as plane stocks and mallets. Southland beech from New Zealand (pinkish brown with silky lustre, rings fairly distinct and rays invisible) is used for similar purposes but is not so durable when exposed.

Birch.—From Europe, including British Isles. Weighs 680 kg/m^3. White to light brown; diffuse porous, pores barely visible; rings and rays barely visible; fairly straight grained; medium texture; strong, tough, not durable; cuts with smooth, bright surface. Used for ply-wood, doors, furniture and motor bodies. Canadian yellow birch (from South-East Canada, Newfoundland and North-East U.S.A.) is light to dark reddish brown and has similar characteristics but is not durable when exposed; used for similar purposes, including aeroplane construction.

Black Bean.—From Australia. Weighs 790 kg/m^3. Dark brown streaked with greyish brown; usually straight grained, but sometimes interlocked, giving a beautiful mottled figure; durable; rather difficult to work. Used for decorative work, including panelling, both solid and as a veneer.

Canary Whitewood (American whitewood).—From U.S.A. Weighs 510 kg/m^3. Yellowish brown with greenish tinge; diffuse porous, pores just visible; rings distinct, rays indistinct; straight grained; easily worked. Used for joinery, plywood. High cost is restricting its use, obeche (see p. 42) being used as a substitute.

Elm (English, Dutch, wych and white elm).—English, Dutch and wych elm are from the British Isles: white elm is from Eastern Canada and the U.S.A. Weighs 560 (English and Dutch), 690 (wych) and 630 (white) kg/m³. English is reddish brown (wych is paler with green streaks) ; ring porous, large pores distinct ; rings and rays distinct ; irregular grain producing attractive wavy figure and uneven texture ; tough ; difficult to work ; durable under water ; wych is strongest ; white elm is usually straight grained and not durable when exposed. Used for weather-boarding, furniture, piles and flooring (*white* only).

Gurjun (apitong, kanyin, keruing and yang).—From the Andaman Islands, Burma, Ceylon, Malaya, Philippine Islands, etc. Weighs 740 Kg/m³. Red to dull greyish brown; straight and interlocked grain; not easy to work ; hard and durable. Used for general constructional work, flooring, bridge decking and wagon building ; also as a substitute for teak (p. 43).

Iroko (odum, African teak and mvule).—From East and West Africa. Weighs 660 kg/m³. Light to dark brown ; interlocked grain producing ribbon figure ; coarse but even texture ; strong and very durable. Used for superior joinery (doors, windows, staircases, flooring, panelling), furniture. Substitute for teak (p. 43).

Lauan, Red and White.—From the Philippine Islands. Weighs 560 kg/m³. Pale to dark reddish brown ; diffuse porous with distinct pores and white chalky resin ducts ; straight and irregular grain producing roe or stripe figure. Allied to mahoganies (see below) for which they are used as substitutes.

Mahogany.—There are several varieties including (1) African, (2) Cuban and (3) Honduras.

1. *African Mahogany* (Accra, Benin, Duala, Cape Lopez and Lagos mahogany).—From West Africa and Uganda. Weighs from 480 to 720 kg/m³. Light pinkish brown to deep red ; diffuse porous, pores distinct with gum deposits ; rings not visible, large rays just visible ; straight and interlocked grain producing roe and striped figure ; moderately durable. Used for good-class joinery, including panelling, veneers, plywood, furniture and similar decorative work.

2. *Cuban Mahogany* (Spanish, West Indian, Porto Rico and Jamaican mahogany).—From the West Indies. Weighs from 640 to 800 kg/m³. Rich reddish brown ; diffuse porous, distinct pores often containing white deposits ; straight, irregular, interlocked and wavy grain producing variety of handsome figure such as blister, roe, stripe and fiddle-back ; ripple marks may be present but not so distinct as Honduras mahogany ; fine texture ; strong ; shrinks and warps little ; difficult to work. Is a true mahogany. Used for superior joinery and decorative work such as panelling, veneers and furniture. Most valuable of the mahoganies, but very expensive and more difficult to obtain.

3. *Honduras Mahogany* (baywood and Central American mahogany).—From British Honduras, etc., Central America, Brazil and Peru. Weighs 550 kg/m³. Similar to Cuban mahogany but colour usually lighter and texture not so fine ; ripple marks distinct ; dark-coloured gum deposits in pores common, white deposits rare ; strong, durable, works easily.

Used for high-class joinery, including panelling, veneers, furniture and similar decorative work.

An excellent substitute for the mahoganies is makoré (cherry mahogany), obtained from West Africa ; it is pale pinkish brown to purplish brown, has a straight and interlocked grain producing mottled figure with occasional dark veins.

Oak.—There are several species of this important timber, including (1) English, (2) Austrian, (3) American Red and White and (4) Japanese.

1. *English Oak* (pedunculate).—Weighs 690 to 850 kg/m³. Heartwood light yellow-brown to deep warm brown, sapwood lighter ; ring porous, spring wood pores distinct ; rings distinct, very distinct broad rays give beautiful " silver grain " effect when rift-sawn ; very durable, tough and strong. Best of species. Used for decorative and superior joinery (including panelling, veneers, plywood, doors, windows and floor boards) and furniture ; carpentry, such as open roofs, exposed floor beams, half-timbered work, fencing and gates. Supply limited.

2. *Austrian Oak* (wainscot).—Is straighter grained than English and therefore slightly less distinctive figure.

Russian and *Polish Oak* are stronger than Austrian oak, but not so well figured ; easier to work and cheaper. *Durmast oak* (France) is less strong and durable than English.

3. *American Red and White Oak.*—From Eastern Canada and U.S.A. *White oak* is somewhat similar to English and preferred to *red oak*, which is usually coarser and inferior. Used for similar purposes to English oak, but for inferior work.

Silky Oak, from Australia, is not a true oak, but is similar to American red oak and has the characteristic " silver grain " figure resembling true oak (hence the name). *Tasmanian oak* (mountain ash and Victorian oak) is light brown and resembles American or plain-sawn English oak, but lacks the " silver grain " characteristic and is not durable when exposed.

4. *Japanese Oak* is lighter than Austrian oak (light brown tinged with grey rather than red), not so pronounced " silver grain " and not so strong ; very even textured ; works easily to smooth finish. More suitable for internal work (such as panelling) than for external constructional work.

Obeche.—From West Africa. Weighs 380 kg/m³. White to pale straw ; interlocked grain producing striped figure ; coarse even texture ; easily worked ; not durable. Used for general joinery, plywood, blockboards. Substitute for Canary whitewood (see p. 40).

Rock Maple (bird's-eye, blister, curly, fiddle-back, hard, sugar and white maple).—From South-East Canada and North-East U.S.A. Weighs 740 kg/m³. Light yellowish brown ; rings distinct as dark lines, rays distinct ; straight, irregular and wavy grain producing bird's-eye, blister and fiddle-back figure ; dense, tough, hard, strong, not durable ; difficult to work. Used for flooring, stair treads, panelling, veneers and furniture.

Sapele (sapele mahogany).—From East and West Africa. Weighs 640 kg/m³. Dark reddish or purplish brown ; interlocked and wavy grain producing attractive blister, roe, stripe and fiddle-back figure ; scented ; very hard and strong ; moderately durable ; not easy to work ; included amongst the mahoganies. Used for superior decorative work, as for panelling, veneers, interior fittings and furniture.

Seraya, Red and White.—From North Borneo. Weighs 560 kg/m³. Similar to lauan (p. 41) and used as substitutes for mahoganies (p. 41).

Sweet Chestnut (Spanish chestnut).—From Europe, including the British Isles. Weighs 560 kg/m³. Light brown heartwood, sapwood lighter ; ring porous ; resembling oak, but rays not visible and therefore silver grain figure is absent ; splits readily ; subject to heart shake. Used for fencing, gates, piles : figured timber used for decorative work. *American chestnut* is similar to, but coarser than, sweet chestnut and is used for similar purposes.

Sycamore (great maple and plane).—From the British Isles. Weighs 630 kg/m³. White or yellowish white ; distinct rings and rays ; straight and wavy grain producing attractive rippled figure ; fine, lustrous texture ; strong, not durable ; works fairly easily. Used for decorative work, as for panelling, choice veneers, table tops, dairy appliances. Supplies are limited.

Teak.—From Burma, Java and Siam. Weighs 660 kg/m³. Golden brown, occasionally with dark markings or flecks ; ring porous, rings and rays indistinct ; straight grained ; not easily worked (saws, cutters, etc., being dulled) ; strong, hard, very durable ; fire-resistant. Used for superior joinery (as for doors, windows, stair treads, shop fronts, veneers and plywood) and carpentry, marine work. *Rhodesian teak* (not a true teak) is similar and used for high-class flooring.

Walnut.—The several varieties include (1) walnut, (2) African walnut, (3) American black walnut, and (4) Queensland walnut.

1. *Walnut* (English, European, Black Sea, French, Circassian and Italian walnut).—Grown in Europe, including the British Isles. Weighs 660 kg/m³. Variable in colour, irregular dark veins on a greyish brown background producing beautiful figure ; finely figured burrs and crotches ; hard, tough, strong, moderately durable, fine texture. Used for superior decorative work, including panelling and furniture ; burrs and crotches highly valued for veneers.

2. *African Walnut* (Benin and Nigerian walnut).—From West Africa. Weighs 560 kg/m³. Yellowish brown background with dark markings (due to gum veins) ; interlocked grain producing ribbon or stripe figure. Not a true walnut. Used for superior decorative work and joinery.

3. *American Black Walnut.*—From U.S.A. Weighs 700 kg/m³. Characteristics similar to English walnut, but darker and more uniform in colour, and used for similar purposes.

4. *Queensland Walnut* (Australian walnut).—From Australia. Weighs 740 kg/m³. Light or pinkish brown to dark brown, with vari-coloured markings ; interlocked and wavy grain producing a broken striped figure ; difficult to work (dulls tools). Used for similar purposes as English walnut.

CHAPTER TWO

PLYWOOD

Manufacture, characteristics and uses of plywood, laminboards,
blockboards, battenboards, composite boards and wall boards.
Plywood box beams.

PLYWOOD is a most important building material and the woodworker should
have some knowledge of its manufacture and characteristics.

Plywood is also known as *reconstructed wood* or *laminated wood*.   It
is a compound wood made up of several thin layers called *plies* or *veneers* ;
these are glued together under pressure and arranged so that the *grain
of one ply is at right angles to the grain of an adjacent ply or plies*.

A board or sheet of plywood usually consists of an odd number of

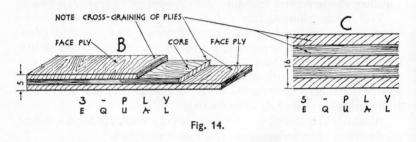

Fig. 14.

plies, *i.e.*, " 3-ply ", " 5-ply ", " 7-ply ", etc.   *Multi-ply boards* are those
which have more than three layers ; 5-ply boards are shown at c, Fig. 14
and D, Fig. 15, and the latter figure shows a 7-ply board at E.

A 3-ply board consists of a middle *core* and two outer or *face plies*.
As indicated these plies are *cross-grained*, *i.e.*, the grain of the core is at
right angles to that of each of the face plies.   The thickness of the plies
varies ; a 5 mm thick 3-ply board will consist of three plies of equal thickness
and is an example of an equal ply board (see B, Fig. 14) ; a 10 mm thick
board, with the same number of plies, has a 6 mm core and two 2 mm face
veneers and is known as a *stout heart* board.   An example of a stout heart
5-ply board is shown at D, Fig. 15 ; this 19 mm thick board has two 2·5-mm
thick face plies, a 6-mm core and two 4-mm intermediate layers or *cross
bandings*.   Examples of equal ply boards are shown at c, Fig. 14 and E,
Fig. 15.

There is a big variation in the sizes of boards.   When specifying sizes the
first dimension quoted is the measurement parallel to the grain of the face
veneer.   Examples of standard or stock sizes of one softwood (Douglas fir)
and two hardwood (birch and mahogany) plywood boards are listed in
Fig. 16.

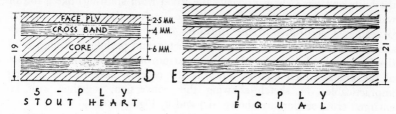

Fig. 15.

STANDARD SIZES OF PLYWOOD BOARDS

Timber	Thickness mm	No. of plies	Length mm	Width mm
Fir	6·5, 9·5, 12·5 12·5, 16, 19 22, 25·5, 32	3 5 7	} 2440 3050	} 1220
Birch	3, 5, 6·5, 9·5, 12·5, 16, 19 22, 24, 25·5	3 5 7	1220 1525 1830 2135 2440 3050 3660	1220 1525 1830 } 1220
Mahogany	as for birch		2440 1525 and door sizes	1220

FIG. 16

MANUFACTURE OF PLYWOOD.—The various processes are : (1) Preparation of logs, (2) conversion, (3) trimming, (4) drying, (5) gluing, (6) pressing, (7) re-drying and (8) finishing.

1. *Preparation of Logs.*—The trunks or boles (see A, Fig. 17) of trees are cut into logs (p. 16) in 250 cm lengths (or according to the size of the converting machine), the bark is removed by hand or machine, hard knots are cut out and any irregularities are removed. Some logs, such as those of beech, gaboon and oak, must be steamed or boiled to soften the fibres ; this takes place in large covered-in concrete pits in which the logs are kept submerged in hot water for two or more days.

2. *Conversion.*—Most logs are converted (p. 16) into veneers by the *rotary cutting method* in which a *rotary veneer cutter*, commonly known as a *peeler*, is employed. A peeler is a powerful lathe with a long sharp knife and a *pressure bar* above it. A prepared log is securely fixed horizontally to the machine by " chucks " which penetrate the timber centrally at the ends. The log is revolved and a continuous ribbon of veneer, uniform in thickness, is cut by the knife ; the veneer passes between

the knife and bar (which prevents the wood from splitting) like a roll of wall-paper being unrolled. The distance between the knife and the bar is regulated according to the thickness of veneer required. The logs should not be less than 250 mm in diameter, and peeling continues until the diameter has been reduced to about 150 mm or until the size and number of the knots (which increase towards the centre) make further peeling impracticable. A sketch showing this conversion, together with an enlarged cross-section are shown at J and K, Fig. 17.

Whilst more than 90 per cent. of veneers are cut by the above method, the resulting figure (with few exceptions, such as that of birch and Queensland walnut) is not very attractive. Modifications of rotary cutting may be employed if highly decorative veneers (as from Cuban and Honduras mahogany, sapele and oak) are required. The *half-round* or *stay-log cutting* method is one alternative; a log is divided longitudinally by a circular saw (Fig. 1 in " Joinery ") and a half log is secured to the peeler with its sawn face against the long knife; as the timber rotates and descends on to the knife a series of separate veneers is produced. *Slicing* is another alternative method to rotary cutting; in one machine (called a *horizontal veneer slicer*) a half log is fixed, with its sawn face uppermost, and a frame having a wide knife with pressure bar is forced over it; the veneer thus cut passes between the knife and pressure bar, the frame is returned to its original position, the timber is automatically raised by an amount equal to the thickness of the veneer and the process is repeated. Burls, crotches and stumps (see A, Fig. 17) are often sliced, especially they are of richly figured timbers. Those with well developed rays as oak) are sometimes radially sliced, *i.e.*, the logs are first quar and each quarter is placed at an angle on the slicer.

3. *Trimming.*—The continuous veneer, as it leaves the peeler, is eit wound on to a spindle which is afterwards taken to the trimming machine or it proceeds along a table to the machine. The latter, called a *clipper*, consists of a long knife which slides vertically in a frame fixed above and across the table. As the veneer passes under the frame, it is cut transversely into widths on each descent of the knife. Patches showing defects, such as splits and large dead knots, are cut out.

4. *Drying.*—The veneers are now dried to the desired moisture content, which varies from 4 to 10 per cent. There are several types of driers. In one the veneers are admitted at one end and proceed at the desired speed between rollers through a heated chamber.

5. *Gluing or Cementing.*—The veneers are carefully inspected, those suitable for face veneers are kept separate from those of lesser quality used as cores. Defective sheets, such as those with pitch pockets and large loose knots, are sent to be repaired or patched; such defects are punched out, replaced by sound patches of the same shape and size having glued edges, and secured under pressure.

The sound veneers are taken to the *glue-spreader*. This machine has a pair of rollers which dip as they rotate into troughs containing the glue.

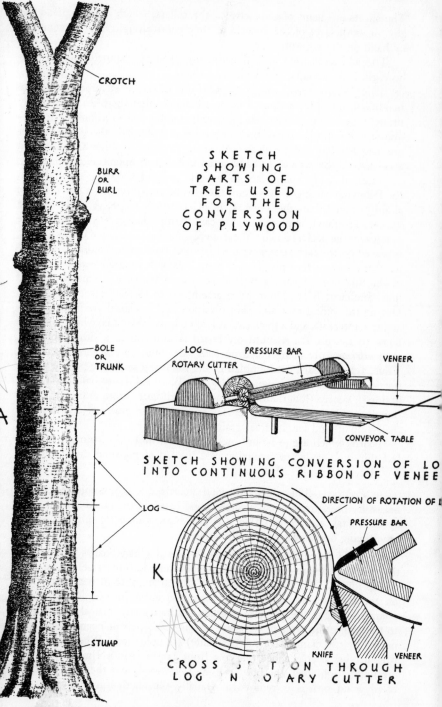

SKETCH SHOWING PARTS OF TREE USED FOR THE CONVERSION OF PLYWOOD

CROTCH

BURR OR BURL

BOLE OR TRUNK

LOG — PRESSURE BAR
ROTARY CUTTER
VENEER
CONVEYOR TABLE

J

SKETCH SHOWING CONVERSION OF LOG INTO CONTINUOUS RIBBON OF VENEER

LOG

STUMP

DIRECTION OF ROTATION OF LOG
PRESSURE BAR

K

KNIFE   VENEER

CROSS SECTION THROUGH LOG AND ROTARY CUTTER

Fig. 17.

The sheets are hand-placed between the rollers, which evenly spread the glue on both faces of the veneers as they pass through, and are removed by hand on the opposite side.

The adhesives used are (a) resin, (b) casein, (c) animal and (d) soya bean glues or cements. See Chapter Eleven

The gluing or cementing process is one of the most important in the manufacture of plywood, as it is most essential that the plies are strongly united together and must remain so when subjected to atmospheric conditions. As stated above, the veneers passed through the glue-spreader are coated on both sides with the adhesive. On emerging they are assembled according to the type of press adopted in the next operation (see below).

Thus, if 3-ply boards are to be cold-pressed they are assembled in the following manner : A 75 mm thick wood board (called a *caul*), of size slightly larger than the veneers, is placed on a bogie. A sheet of face veneer, not glued, is placed on the caul with its *face-side* (the top surface as it leaves the peeler) down ; the core ply, having been glued on both sides, is placed on the face veneer with its grain at *right angles* to that of the face ply ; the second face ply, face-side up and with its grain *parallel* to that of the first face veneer, is carefully laid over the core. This is repeated until sufficient boards have been assembled to form 900 mm thick pile. During this piling process a thin plywood caul is placed at approximately 305 mm intervals, and a thick caul is laid on top of the batch which is at once taken to the press. A multi-ply board is built up in a similar manner, each *alternate* ply being glued on *both* sides and *cross-grained* assembled. Thus, a 7-ply board is assembled in the following sequence : Face veneer (face-side down and longitudinal-grained), glued cross-banding (cross-grained), veneer (longitudinal-grained), glued core (cross-grained), veneer (longitudinal-grained), glued cross-banding (cross-grained) and face ply (face-side up and longitudinal-grained).

Boards which are to be hot pressed are assembled in a similar manner, but thin zinc cauls are generally used instead of plywood cauls, and two are placed between each board.

6. *Pressing.*—The glued plywood boards are now subjected to the necessary pressure to effect a sound bond between the glued sheets. This operation takes place in either a (a) cold press or a (b) hot press, both types being operated by hydraulic power.

(a) *Cold Pressing.*—The assembled pile of glued boards is placed in the centre of the press upon several steel beams laid transversely at intervals on the lower plate of the machine ; a similar number of beams is placed on the top caul and immediately over the lower beams. The press is operated and, as the pressure is gradually increased, vertical clamps fixed to the projecting ends of the steel beams (each pair of beams being connected by two clamps—one at each side) are tightened. When the desired pressure has been reached (varying from 555 to 1040 kN/m²), the clamps are finally tightened, the bottom plate is lowered, and the clamped batch of plywood boards is removed. Usually the clamps are removed after eight hours' application. Plywood boards which have been either casein glued or soya bean glued are cold pressed.

(b) *Hot Pressing.*—Resin cemented and animal glued plywood board must be hot pressed to ensure a strong bond between the plies; casein glued boards are also sometimes hot pressed. A hot press consists of a bottom metal table and twelve or more hollow steel plates, spaced at regular intervals, and to which steam is admitted. Not more than two assembled boards, with their zinc cauls, are placed between each pair of plates. The press is closed, *i.e.*, the bottom table is raised, and this in turn lifts the plates and reduces the space between them; the pressure is gradually increased until the boards are subjected to that required (between 1040 and 2080 kN/m²). Meanwhile the heat from the plates is transmitted to the plywood to effect a strong bond between the glued surfaces of the plies. The pressure is maintained for several minutes, this " bonding time " being variable according to the glue employed, nature of the wood, thickness of the boards, etc. The temperature also varies from 60° C (for animal glued boards) to at least 165° C (for certain resin cemented boards). On removal from the press, the plyboards are sticked (p. oo) and a steel beam is placed on top of the pile.

7. *Re-drying.*—During the gluing and hot-pressure processes the plywood boards absorb moisture, and the moisture content must therefore be reduced. The sticked piles from the hot press are taken to a re-drying chamber and the moisture content is reduced to usually 8 per cent. Cold pressed boards, after the clamps have been removed, are sticked and the piles re-dried.

8. *Finishing.*—The re-dried boards have their edges trimmed as they are accurately sawn to a desired length and width (see p. 45). Finally, the boards are given a smooth finish to both sides by being machine planed and sanded.[1]

MERITS OF PLYWOOD.—1. The shrinkage and expansion of best grade plywood is almost negligible. This is due to its cross-grained construction. As stated on p. 18, the maximum shrinkage of timbers occurs in the tangential direction, and longitudinal shrinkage is very small indeed. Therefore a single sheet of veneer will shrink or expand in its width but will move very little in its length. Such movement will, however, be considerably restricted when the sheet forms part of a cross-grained plywood board. Thus, any tendency for the core of a 3-ply board to " move " (shrink or swell) in its width is restrained by the longitudinal fibres of the two face veneers to which the core is glued. Likewise, any tangential movement of the face plies will be restrained by the longitudinal grain of the core.

2. Plywood is stronger than ordinary timber of the same area and thickness. This is also due to its cross-grained construction.

3. It does not readily split when nailed near to its edges.

[1] These machines are described in " Joinery."

suitable for wall panelling when, if desired, framing can be dispensed with. Formerly, when conversion was limited to the methods described on p. 17, the maximum width of an unjointed panel depended upon the diameter of the log, so extra panel framing was necessary.

5. The modern tendency of using thin veneers, instead of relatively thick panels and framing, for panelling furniture, etc., has resulted in the economical employment of rare and valuable timbers.

The unattractive figure of most timbers when rotary-cut is the only demerit of good quality plywood.

USES OF PLYWOOD.—Its various uses include (*a*) covering or panelling walls, floors, ceilings and partitions ; (*b*) doors ; (*c*) stairs (balustrades) ; (*d*) furniture ; (*e*) temporary carpentry, such as formwork for concrete structures ; (*f*) coach, bus, motor car, etc., construction, and (*g*) box beams. The cheaper varieties are used for boxes, chests, barrels, etc.

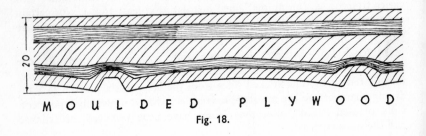

20

M O U L D E D   P L Y W O O D

Fig. 18.

MOULDED PLYWOOD.—This is a development of ordinary plywood and is used for wall panelling. It is moulded on one face to a wide range of patterns, one of which is shown in Fig. 18. It is made as already described except that in the press a metal or wood mould or *form*, having a surface shaped to the reverse of that required on the board, is placed above it, and the moulded face is formed when pressure is applied.

METAL-FACED PLYWOOD.—This is a plywood board covered on one or both sides with thin sheets of metal such as aluminium, bronze, nickel, stainless steel, etc. Fig. 19 shows a portion of a 3-ply board metal-faced on both sides. The metal is bonded to the plywood by special cements. It is also obtainable as boards with narrow vertical, etc., metal bands, screwed to or inlaid flush

VENEER OF LIGHT GAUGED METAL SUCH AS
STAINLESS STEEL, BRONZE & NICKEL ALLOYS, ETC.

SEALED END   6MM. THICK PLYWOOD BOARD

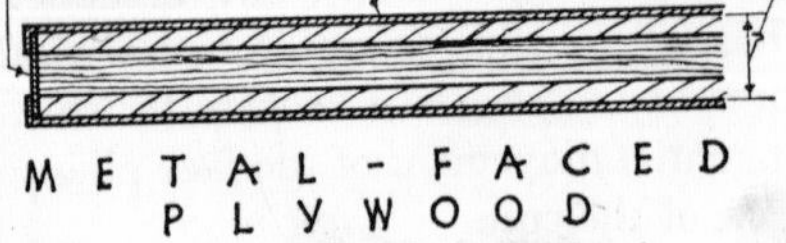

M E T A L - F A C E D
P L Y W O O D
Fig. 19.

with the surface as required, in addition to metal angles at the edges. It is used for wall panelling (single metal-faced), doors, partitions, etc.

LAMINBOARDS OR LAMINATED BOARDS.—These are another development of plywood and are in big demand for floors, partitions, panelling,

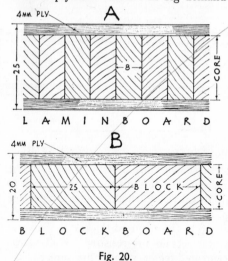

furniture, etc. A portion is
shown at A, Fig. 20. As
shown, the core is built up of
thin strips or laminæ not
exceeding 8 mm wide and
glued between two or more
plies. A common size of
laminboard is 1800 long, 3600
mm wide and 13 to 50 mm
thick.

BLOCKBOARDS are similar to
laminboards except that the
core is built up with blocks of
wood not exceeding 25 mm
wide, as shown at B, Fig. 20.

Fig. 20.

They are cheaper than laminboards and are used for similar purposes,
although laminboards are preferred for first-class richly veneered work.

COMPOSITE BOARDS (Fig. 21).—These are made of several wood plies
with one or two layers of asbestos fibre (see below) or similar material which
is non-combustible and sound insulating. Such a board, having two insulating layers, is shown in Fig. 21. It is used for covering ceilings, walls and
partitions when a reduction of sound transmission (pp. 171–176) and
relative non-combustibility are essential requirements.

Fig. 21.

BATTENBOARDS (see Fig. 23).—With the exception of the core, which is
formed with edged-glued battens not exceeding 75 mm in width, these are
similar to blockboards. They are not much used in this country, blockboards being preferred.

WALL BOARDS.—These are sheets of asbestos-cement (2·4 m by 2·4 m
by 6 mm) or wood fibre (shavings) or wood pulp (1·8 to 4 m long; 0·9 to
1·2 m wide and 10 to 25 mm thick). They are partially sound-proof (see
p. 171) and are used for covering walls, partitions and ceilings and as de-

scribed on pp. 172–176. Asbestos-cement is composed of asbestos (a silky
fibrous mineral obtained chiefly from South Africa) and Portland cement

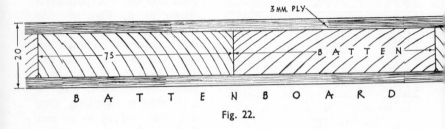

Fig. 22.

(made in this country from chalk and clay); the asbestos rock is dried,
broken, crushed, passed through a screen, mixed with water and the
cement (1 part asbestos to 7 parts cement), the mixture is transferred to
another machine (a revolving cylindrical sieve, which removes the excess
water) from which the material passes to a large forming cylinder where a
sheet of asbestos is gradually built up to the required thickness; the sheet
is finally pressed. Asbestos-cement sheets are fire-resisting. Wall boards
of wood fibre have the shavings cemented together under pressure.

PLYWOOD BOX BEAMS.—These are used to make timber beams for large
spans, they incorporate top and bottom solid timber chords to which
plywood webs are glued and nailed down each side.

The part elevation at N, Fig. 23 shows a typical example for a 9 m span ;
such a beam, placed at 1·5 m centres would support a normal domestic floor
load. Small timber floor bridging joists would span between the beams and
rest on timber fillets screwed and glued to the plywood webs.

The box beam shown has 125 mm by 75 mm top and bottom chords
with 16 mm plywood webs glued and pinned to them. Vertical braces of
the same section as the chords are placed within the web at 600 mm intervals.
The construction is shown clearly at O, P and Q ; those at O and P also show
the connection to a timber column made from two 100 mm by 75 mm
timbers glued and pinned together. The fastening is made by making the
plywood webs project so that they lap the sides of the column where they
are nailed. The ends of the beams are strengthened by cranked metal plates
made from 75 mm by 10 mm steel which pass down the brace and are
cranked over for a distance of 100 mm.

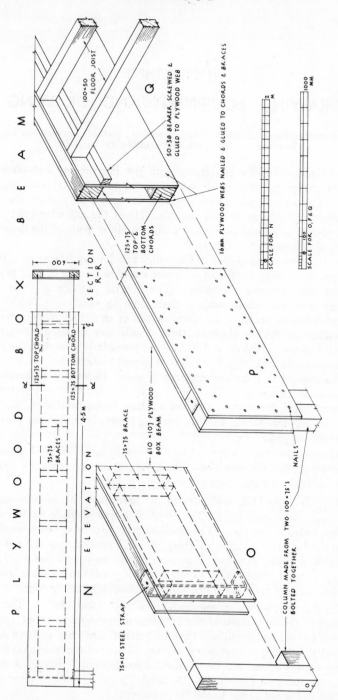

Fig. 23.

CHAPTER THREE

DRAWING EQUIPMENT AND DRAWING

Instruments, paper ; hints on draughtsmanship ; methods of representation ;
lettering, sketching, inking-in, tracing and colouring.

THE student should be conversant with the instruments and materials used in the preparation of drawings related to carpentry and joinery. It should also be the aim of the beginner to reach a satisfactory standard of draughtsmanship as early as possible. Hence the following brief notes which should be considered before a close study is made of the details of construction appearing in this book.

Students attending classes in Carpentry and Joinery (theory as distinct from practical) will be required to devote a certain amount of time to scale drawing and freehand sketching. The general practice at each class meeting is to commence with a clean drawing sheet upon which the student draws to scale, or sketches, details of construction prepared by the teacher on the blackboard ; these details may be supplemented by examples in the textbook, and if it is not possible to complete the sheet in class the student is required to do so as part homework.

The student is therefore required to equip himself with certain instruments and materials. The following are the minimum requirements : Drawing board, tee-square, two set squares, scale, protractor, compass with pencil point, pencil, rubber, drawing paper, four drawing pins and a notebook or pad.

DRAWING BOARD.—A type of board quite suitable for the beginner is made of 3-ply (see Chapter Two). A convenient size is 605 mm by 430 mm as this will take A2 size sheet of paper (see p. 55).

TEE-SQUARE.—This consists of a hardwood blade screwed at right angles to a short head. The standard length of blade for the above drawing board is 605 mm. The cheapest type, usually of beechwood, has a thin parallel blade ; it is satisfactory, although easily damaged if not carefully looked after. A better, but more costly, tee-square has a polished mahogany taper blade with a bevelled (sloping) edge in which is inserted a narrow strip of celluloid which forms a transparent ruling edge.

SET SQUARES.—Two are required, i.e., a " 45° " and " 60° ". The sides adjacent to the right angle of the 45° set square should be about 150 mm and the 60° set square should have its vertical side (that adjacent to the 90° and 30° angles) at least 200 mm long. Those of transparent celluloid, having square edges, are best ; the " open pattern " (having a central triangular space) is preferred to the solid " pattern," as it is more easily handled and is not so apt to soil the drawings.

An adjustable set square is a useful but not an essential instrument ; its sloping edge can be set to any desired angle.

SCALE.—A scale 300 mm long, of oval cross section and made of box-wood is recommended. It has eight scales, *i.e.*, 1 : 10, 1 : 100, 1 : 20 and 1 : 200 on one side, and 1 : 5, 1 : 50, 1 : 1250 and 1 : 2500 on the other. The edge of the scale containing the 1 : 10 and 1 : 100 scales is calibrated in millimetres and so can be used for full size (1 : 1) drawings. A 150 mm pattern, having the same number of scales, is sometimes preferred, as it is handier to carry in the pocket. It is also made of ivory, boxwood with white celluloid edges, and ebonite (black with white divisions). Cardboard scales are not recommended.

PROTRACTOR.—This may be of the 150 mm rectangular type or, pre-ferably, of the semicircular transparent celluloid pattern of 100 or 125 mm diameter which is divided into degrees. This is not necessary if an adjustable set square (see above) is available.

COMPASS.—The student should take special care in selecting this; As a rule, very cheap compasses quickly develop defects and are responsible for inaccurate or untidy draughtsmanship. Whilst a pencil compass is all that he may require at first, he may later wish to acquire additional instruments as he advances in his studies. Several makers now produce sets of reliable instruments at a reasonable price specially to meet the requirements of students. Such instruments are standard and therefore any replacements of damaged or missing parts can be readily obtained at a minimum of cost. Instruments should be accommodated in a case for convenience and as a protection.

A case containing a 140 mm compass having a pencil socket, loose dividing point and a refill case for holding spare leads will meet the require-ments of the beginner. When not being required for use as a pencil compass, the lead in the socket may be replaced by the dividing point for pricking off lengths or dimensions, etc.

Alternatively, a larger set which will serve the needs of the student for several years consists of a case containing the following : (*a*) 125 mm compass with pencil, pen and lengthening bar fittings, and a key for tighten-ing joints ; these fittings are interchangeable, thus the pencil is replaced by the pen fitting when curves are to be inked-in, and the bar is introduced when larger circles or arcs are to be described ; (*b*) 125 mm divider ; (*c*) 90 mm spring-bow pencil (for small circles) ; (*d*) 90 mm spring-bow pen (for inking-in small circles) ; (*e*) 125 mm drawing pen (for ruling lines in ink), and (*f*) refill case with spare leads.

PENCIL.—A good quality H. or H.B. pencil is recommended (see p. 57).

RUBBER.—This must be soft and free from grit, otherwise it will damage the surface when applied to drawing paper. An ink rubber is useful if, at a later stage of his training, the student inks-in his drawings or produces tracings.

DRAWING PAPER, ETC.—This is required for both class and homework. Good quality paper is essential. As stated on p. 54, the size of the sheets recommended for first-year students is A2 size (594 mm by 420 mm).

Four drawing pins are needed for fixing the paper to the board, and some should be kept in reserve. A *sharp* penknife is also essential (see next page).

NOTEBOOKS AND PADS.—Either a notebook or pad is essential. The paper, which should be plain, should be of good quality and preferably of A4 size (210 mm by 297 mm). Ruled paper should not be used, as students from the outset should learn to sketch unaided by printed lines (see p. 68). If a pad is used, the sheets should be filed methodically in a suitable folio or folder.

Finally, instruments should be treated with care. Good work cannot be expected if the edges of tee-squares, set squares and scales are chipped or otherwise damaged, and if the joints of compasses become permanently loose or the legs bent.

HINTS ON DRAUGHTSMANSHIP, ETC.—A sheet of paper is pinned to the drawing board in the following manner : A drawing pin is inserted in the top left corner of the sheet and board. The sheet is " squared " (the tee-square being used to ensure that the top edge of the paper is parallel to that of the board), drawn taut by hand-pressing diagonally towards the bottom right-hand corner at which a second pin is inserted, stretched from the centre towards the top right corner which is then pinned, and tightened towards the remaining corner in which a pin is pressed home.

Most draughtsmen prefer to work on drawing boards which have sloping surfaces. Some boards are provided with shaped bearers which give the required inclination. Otherwise a board may be tilted by placing under its top edge a piece of wood (of the desired thickness and sufficient length), an old book, etc.

All horizontal lines should be drawn aided by the tee-square. The head should be pressed against the left edge of the board by the left hand, the tee-square is slid up or down until the blade is in the desired position, and this pressure should be maintained until the line has been drawn. The pencil should be held in contact with the *bottom* of the ruling edge of the blade as the stroke is made from left to right. Students are apt to acquire the bad habit of moving the tee-square with the left hand at the middle of the blade and keeping the hand in this position whilst drawing the line ; this often results in the lines not being " square ".

Vertical lines should be drawn with the assistance of a set square. The tee-square is placed in the desired position with its head pressed against the left edge of the board, as described above. The position is maintained as the left hand traverses and presses on the blade. The set square is placed with its base on the upper edge of the blade and its vertical side in the required position. Both tee-square and set square are held by the left hand whilst the vertical line is drawn with the pencil contacting the bottom of the vertical edge. Set squares are also used for drawing lines at 45°, 30°, 60°, etc. Long vertical lines, such as borders, may be drawn with the tee-square.

The scale should never be used as a ruler, as its edge will become damaged by the action of the pencil—or pen—against it.

Each line should be of uniform thickness. Students should practise drawing lines of varying thickness against the edges of the squares until they can with certainty produce lines of the desired strength.

It cannot be too strongly emphasized that *the quality of draughtsmanship is largely dependent upon the condition of the pencil, and blunt pencils are a common cause of drawings which are inaccurate and of unsatisfactory appearance.* Hence the pencil should be sharpened to a long (at least 20 mm with 6 mm lead exposure) and tapering point, and *it must be maintained in this condition by the frequent application of the knife.* Whilst drawing lines, revolving the pencil after each stroke will help to maintain the point. A blunt pencil is often due to a blunt knife. A stumpy pointed pencil must not be used as it blunts quickly and can only be applied against the bottom edge of the tee-square with difficulty (see p. 56). Short pieces of pencil, unless fitted in a suitable holder, should not be used as they are difficult to control.

Before commencing drawing, the student should see that the tee-square, set squares and scale are clean, otherwise the sheet will quickly become soiled. A clean duster or waste piece of plain paper should also be used for keeping this equipment clean as the drawing proceeds.

Until the student becomes reasonably efficient as a draughtsman, he should at first apply the pencil lightly to the paper. Apart from the difficulty in erasing firm lines drawn in error, he will find that heavy lines (especially when drawn with a soft H.B. pencil) will quickly become smudged by the constant action of the tee-square and set square over them. These instruments also collect the spare lead and transmit it over the paper to produce a dirty sheet. The latter should be cleaned down, by a careful application of the rubber, after the whole drawing has been completed with the lightly drawn details. The drawing is then " lined-in ". *Before this is attempted the pencil should be sharpened and the squares cleaned.* He should commence lining-in at the top of the paper, working from left to right and gradually towards the bottom. Any printed matter is added *after* the lining-in has been completed. Some notes on lettering are given on pp. 62–66. When setting out the drawing, sufficient space should be left for any titles and sub-titles. A piece of clean paper below the hand will prevent the drawing from becoming soiled during printing. Finally, the four thick lines of the border are drawn to complete the drawing.

METHODS OF REPRESENTATION.—A student attending a complete course of instruction will be taught in his geometry class several methods of representing on paper certain geometrical solids (such as cubes, prisms, pyramids, etc.) and consideration will there be given to the practical application of solid geometry to building problems. Hence, it is only necessary to make brief reference here to these methods which are :—
(1) Orthographic projection, (2) isometric projection, (3) oblique projection, (4) axonometric projection and (5) perspective.

1. *Orthographic Projection.*—This is the form of representation which is chiefly used. It is, for example, the method adopted for working drawings comprising plans, elevations and sections of buildings.

A *plan* is the view presented when looking vertically down on an object. *Projectors* are perpendiculars ; if these be drawn from the corners of an object to points on a horizontal flat surface (called a *plane*) and if these points be connected by lines, the resulting figure is known as a *projection* and is the plan of the object. If, for example, the object is a building plot, the plan will show its true shape as its boundaries will be of true length and correctly related. A plan of a room will show its true shape and the correct positions, etc. of the door, window(s) and any fireplace. Working drawings include a separate plan of each floor of a building, and the following terms are used : " Basement Plan ", " Ground Floor Plan ", " First Floor Plan ", etc. Examples of plans are shown on pp. 25 and 103.

An *elevation* is the view obtained when looking in a horizontal direction towards the object. It is the projection on its vertical plane. All vertical lines are of true length. Thus, an elevation of a house, when developed from the plan, will show the true shape, dimensions and position in a wall of the door and windows, and will include the roof, chimney-stacks etc. Terms applied to a building, such as " Front Elevation ", " Back Elevation ", " End Elevation ", " North Elevation ", etc., are self explanatory. Examples of elevations are shown in Figs. 113 and 139.

A *section* is the true shape of an object presented after it has been cut through (or assumed to have been divided) by a plane, and the portion between the observer and the plane removed. If the cutting is vertical, a *vertical section* results (see Fig. 77 and A, Fig. 94). That called a *horizontal section* is produced by a horizontal cut; a ground floor, etc. plan of a building (which shows the thickness of the walls, etc.) is really a horizontal section. Sometimes the dividing plane is indicated on the plan or elevation and distinguished by letters, and the section is similarly lettered (see Fig. 24). A further distinction is made between *longitudinal section* (see B, Fig. 120) and *cross* or *transverse section* (see Fig. 121), according to the positions of the actual or imaginary dividing planes. Sometimes

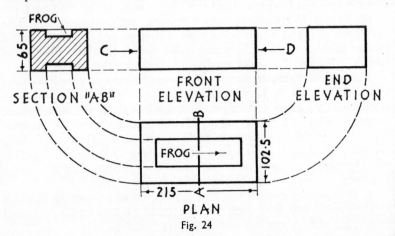

Fig. 24

a view shows an elevation combined with a section ; such is called a *sectional elevation*.

A simple example incorporating a plan, elevations and section is shown in Fig. 24. The object is a pressed brick which has usually two depressions or frogs. The plan is first drawn to scale (see p. 62), the external dimensions being assumed to be 215 mm by 102.5 mm. The frog is then indicated, as shown. The cutting plane AB is drawn. The vertical section (" Section AB ") is next developed and is a cross section ; if the plane had been in the direction of the length, a longitudinal section would have been required. The width of the section is, of course, 102·5 mm and its height is assumed to be 65 mm. The two frogs must be shown. Finally, the cut surface is *cross-hatched*, *i.e.*, diagonal lines (45°) are drawn at close intervals as shown (see p. 68). The front elevation is developed by projecting vertical lines from the plan and horizontal lines from the section. The end elevation is then developed. The correct position of the end elevation in relation to the front elevation must be carefully noted. As shown, the end elevation is that of the brick when viewed in the direction of the arrow C. If viewed in the direction of arrow D, the end view would be shown on the left of the front elevation. Actually, as both ends of the brick are the same, this is not an important matter in this case, but when the end views differ it is essential that they be represented in the correct relative position. In actual practice the broken concentric quadrant curves are not drawn ; they have been indicated in the figure to show the developments more clearly. The frogs could be shown by broken lines in both elevations, but these have been omitted for the sake of simplicity.

2. *Isometric Projection.*—In orthographic projection, as illustrated in Fig. 24, three separate drawings are required to show the plan and the two elevations. When isometric projection is employed, these three views are embodied in a single drawing. Thus, an isometric drawing of the same brick is shown in Fig. 25. A 30° set square is used to draw the length and width and lines parallel to them. In building drawing, the height, length and width are drawn to scale to the actual dimensions. The drawing gives a distorted shape, as the angle between any two adjacent sides is either

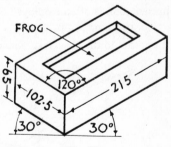

Fig. 25.

120° or 60°, instead of 90°. Nevertheless, this method of representation is most useful and is often employed in either drawing to scale or sketching details of construction. Figs. 33 and 93 are examples of isometric drawings to scale, and an isometric freehand sketch is shown in Fig. 144

3. *Oblique Projection.*—Fig. 26 is a drawing by oblique projection of the pressed brick. One of the vertical faces (the end face in this case) if

assumed to be at right angles to the line of sight and is drawn true to shape and scale. The adjacent vertical face is drawn obliquely, usually with the 45° set square, although any angle may be adopted. In order to reduce the distortion, the dimensions of the inclined lines are sometimes halved. This is not so often used as the isometric method.

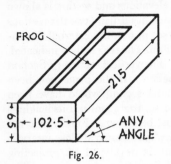

Fig. 26.

4. *Axonometric Projection* (see Fig. 27).—This resembles isometric projection in that both the adjacent vertical faces are inclined. There is, however, an important difference between the two, for, whereas the plan of the isometric drawing is distorted, that axonometrically drawn is of true shape. Observe in Fig. 27 (which is that of the pressed brick) that the sides of the top are at right angles. When presenting an object by this method, the plan is drawn true to scale, set at *any* convenient angle, and the elevations are then projected from it. The angle at which the plan is drawn varies in accordance with its shape and any special requirements—such as an important portion requiring emphasis—the aim being to provide the maximum information. B, Fig. 84 is an axonometric sketch. This method of presentation is often employed.

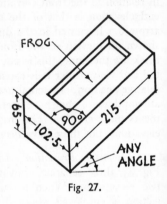

Fig. 27.

5. *Perspective.*—In a perspective drawing the object is represented as it appears to the observer. Hence, to conform to the impression produced

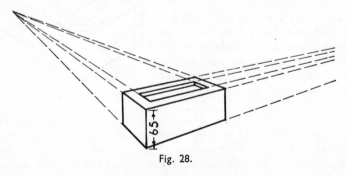

Fig. 28.

upon the eye, horizontal lines which are actually parallel are drawn converging towards a distant point. Also, the heights of objects are shown to diminish, as the distance from the observer increases. The principles of perspective are beyond the scope of this book, but some idea of the features of this form of representation may be obtained by reference to the simple example illustrated in Fig. 28. The horizontal lines converge towards two points, one on either side of the object (a pressed brick), and it will be seen that the only true dimension is the figured height of the arris or edge. Perspective drawings serve a very useful purpose as they are more realistic than geometrical drawings, but correct measurements cannot, of course, be taken from them.

APPLICATION OF THE SCALE.—When a drawing is made the same size as the object it is said to be " drawn to full size ". Whilst it is rarely necessary to draw carpentry details to full size, large scale joinery details are very frequently called for. Thus, for example, full-size details of portions of doors, windows and stairs are required to enable the joiner to set out in pencil on a board the details to full size prior to cutting and shaping the various members detailed.

Most carpentry details are drawn to reduced dimensions, as it is not necessary, nor is it practicable because of their large dimensions, to draw them to full size. For instance, a full-size scale drawing of the portion of the floor, etc. shown at E, Fig. 69 would occupy a space of approximately 900 mm by 700 mm.

The extent of the reduction depends upon the size of the object, the purpose of the drawing, size of sheet, etc. Thus the details at E, G and F, Fig. 73, were originally drawn to a scale of 1 : 10. By this is meant that every 10 mm of actual length of the object is shown on the drawing by 1 mm and every portion of 10 mm by the same fraction of 1 mm. Hence a wall 920 mm long would be represented on paper by a length of 92 mm.

Sometimes details, such as that of an eaves of a roof (Fig. 103), are drawn to " one-fifth full size " (i.e., 1 : 5) and " half full size " (1 : 2). Before a house can be built, a " plan " of it must be submitted, in duplicate, to the local authority for approval. This may be referred to as a " general working drawing " or a " contract plan "; such includes plans of the several floors, elevations and at least one vertical section. It is usually drawn to a scale which is expressed as " 1 : 100 " or " 1 mm to 100 mm. Less frequently a scale of " 1 : 50 " is selected for such a drawing. Quite commonly the elevation of a building, or portion thereof, is drawn to a scale of 1 : 20.

It will be seen from the foregoing that a scale drawing is proportionate to the actual object. This ratio of reduction is called the *representative fraction*, and the scale is often referred to by such a fraction. Thus, the representative fraction of the 1 : 20 scale is $\frac{1}{20}$.

Each drawing must have its scale clearly indicated on it; students should not, therefore, omit this from their class or homework sheets. This may be shown either by (a) a printed statement such as " Scale : 1 : 100 or (b) a portion of the scale, accurately drawn with a sharp pencil, in a convenient position (often at the bottom of the paper) in a manner similar to that shown in Fig. 32 ; if the latter, measurements not figured are scaled from the drawing with the aid of dividers.

DIMENSIONS.—Drawings should be fully and clearly dimensioned. *This is most important.* The figures must be distinct ; special care being taken with printing the figures 3 and 5. Two methods of indicating dimensions are shown in Fig. 31. The dimension lines should be lightly drawn and terminated against short thin lines which are continuous with the ends of the portion of the drawing concerned (see also Fig. 24). Ambiguity arises when the short terminal lines are omitted, as in the example " 1 " immediately below the title of Fig. 30. If a figured dimension does not agree with the scaled dimension (as occasionally happens in practice when an alteration is made to a working drawing), the figures must be taken as correct and either underlined or followed by the letters " N.T.S." (meaning " not to scale "). Figures are again referred to on p. 65.

PRINTING OR LETTERING.—A drawing is not complete without a printed title or heading, sub-title, etc. Good lettering improves a drawing and, conversely, the appearance of a sheet is spoiled by bad lettering. Lettering must be distinct, and the letters must be well spaced and of pleasing form. That used for working drawings, class sheets and homework should be simple and easily formed in a reasonable time. An elaborate style of printing is now rarely called for ; indeed, the practice in many drawing offices is to spend as little time as possible on lettering, and hence it is often produced by means of the stencil ; the latter is a plate with perforated letters and is used with a special pen.

A recommended plain type of lettering which, after a little practice, is easy to do is shown in Fig. 29. The student should study this carefully. Each letter is shown within a square in order that the proportions may be better appreciated and to help comparison. He should note that B, E, F, J, K, L, P, R and S are half letters (occupying half width of square), C, D, G, H, N, T, U, X, Y and Z are three-quarter letters, and A, M, O, Q, V and W are full letters. He should observe especially that O and Q are complete circles and C, D and G are three-quarter circles. Figures are also shown. The sloping or italic is another good style of lettering ; the letters are inclined at an angle of approximately 10° ; most beginners find this more difficult than the upright version.

Many students pay too little attention to printing, with the result that their drawings are spoilt by hurriedly formed letters which are ugly in the extreme. In order to show examples of bad lettering Fig. 30 has been prepared, consisting of letters *reproduced from actual homework sheets.* Unfortunately, these are not by any means isolated examples. Such crudity is to be deprecated. The following few observations are made on these examples as most of the defects will be self-evident.

The first letter B, very popular with students, is too wide (compare with that in Fig. 29) ; such excessive width is also often given to letters E, F, K, P, R and S. On the other hand, C, D, G, O and Q are usually too narrow. Letters with flourishes and curlicues (see especially the first S) cannot be too strongly condemned. Regarding the printing at " 2 ", apart from the badly-spaced ill-formed letters, capitals should not be mixed with the lower case (as " and "). The space between each of the finicky lettered words of the title in " 3 " is excessive. Capital I's must not be dotted as shown in " 4 " ; the first letter of each word should not as shown, be higher than the rest ; it is unnecessary to draw attention to the small n's mixed with the badly shaped capitals and to the wrongly placed apostrophe. The placing of a title between two bold lines, as in

Fig. 29.

" 5 ", is undesirable and a single line only as in, underlining, is less common than formerly ; the pair of short lines at the ends are inexcusable ; if serifs (short strokes as at T and H) are used they should be applied throughout and not occasionally as shown at " 5 ". As stated below, printing should be between *faint* guide lines, otherwise the irregular lettering shown at " 6 " results. Attention has already been drawn to " 1 " (p. 62) which shows a dimension without the required terminal lines   the scribbled figured " 3 " is hardly distinguishable from a 5.

Reference is made on p. 66 to the three ugly arrows shown in Fig. 30 and to the fancy bottom corners of the border.

A few hints on lettering are given in Fig. 31. The height of the main title depends upon the size of the sheet and the space available, but 9 mm letters are suitable for an A2 size sheet. Sub-titles, such as " Plan ",

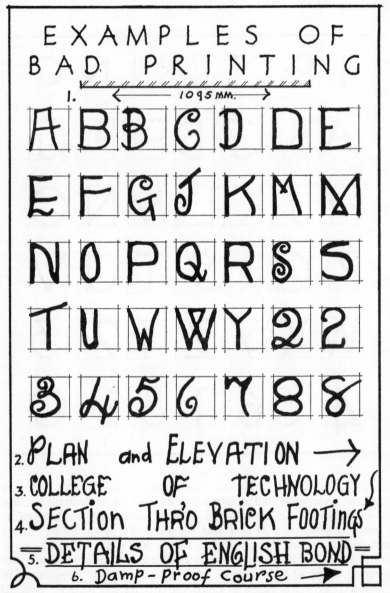

Fig. 30.

" Elevation ", etc., may be from 3 mm to 6 mm high.   General descriptive and brief specification notes are usually of 3 mm to 6 mm letters and figures.

Letters and figures, no matter how small, should be between *very faint* top and bottom guide lines, if untidy and uneven lettering (see " 6 ", Fig. 30) is to be avoided.

Good lettering depends largely upon correct spacing.   Letters in a word must not be placed the same distance apart although the spaces between them should *appear* equal ; otherwise a letter is either given undue prominence, due to an excessive space on one or both sides of it, or a portion of the word may appear unduly crowded.   No mechanical rule can be applied owing to the varying shape and sequence of the letters, but special care should be taken when spacing adjacent letters such as A and V (see " 5 ", " 6 " and " 7 ", Fig. 31), A and W, L and T (see " 8 ", " 9 " and " 10 "), R and I, P and A, O and W, W and Y, etc.   The space between letters must, therefore, be adjusted according to their shapes.   Hence, in order to attain interspaces of apparent equal areas, the distance between adjacent letters having *upright strokes must be increased* (see " 3 "), that between *round letters must be decreased* (see " 4 ") and the distance between a letter having a *vertical stroke and one with a curved stroke* (as " N " and " G " at " 1 ") *must be intermediate.*

When spacing letters forming titles and sub-titles, it is helpful if, after the guide lines have been *faintly* drawn at the desired height apart, points are *lightly* marked off at an equal distance apart and *very faint* short vertical lines are drawn from these points which serve as centre lines for the letters which are then firmly pencilled freehand.   This method is shown at " 2 ", Fig. 31, but with the vertical lines stronger than is normally necessary to meet reproduction requirements ; note that the distance between the words is twice that between the letters.   When the student becomes proficient he may dispense with the vertical lines (which also help to maintain the verticality of the letters) and be guided solely by the scaled off centre points.   The following additional example is provided to make clear this method of spacing.   Suppose the student has completed a sheet of carpentry details and he is about to start lettering it.   Commencing with the title, which is assumed to be " CARPENTRY ", he will faintly draw the horizontal guide lines at 9 mm apart and mark off from the *centre line* of the sheet four equal spaces (say 25 mm) on each side, erect the short faint vertical lines and then proceed to print the letters with half the width of each on both sides of the upright.   Later he will become aware of the need to make slight adjustment in the spacing of the centres, especially when it is relatively small and there is a big variation in the shapes of adjacent letters ; thus, in the title " LETTERING " in Fig. 29 the space between the centres of R and I is slightly less than the average and that between N and G is somewhat greater.

The description in Fig. 31 consists solely of capital letters.   Some prefer to adopt the lower case (small letters).   Horizontal guide lines (which were subsequently removed) were used for the dimension figures shown in the top left corner.

ARROWS.—These should be neatly drawn (preferably aided by the tee-square or set square) and carefully finished with *small* heads (see Fig. 31). A drawing is completely marred by large-headed arrows which sprawl at random over the sheet. Three bad examples are shown in Fig. 30. Incidentally, unlike that shown at " 4 ", Fig. 30, an arrow head should point towards the detail concerned.

BORDER.—This completes the sheet and should preferably be a single heavy line drawn to leave about a 12 mm wide margin. Finicking corners, such as are indicated at the bottom of Fig. 30, should be avoided.

SUMMARY.—In general, each sheet of drawings produced in class or homework should :—

1. Consist of neat and accurate details, drawn lightly at first and subsequently lined in (p. 57) and arranged with the elevation over and projected from the plan, etc. (p. 58).

2. Be carefully set out to ensure a reasonable balance, *i.e.*, details should be suitably spaced to ensure a uniform covering of the sheet and the avoidance of local overcrowding and large blank spaces.

3. Be cross-hatched and lettered after the scale drawing has been completed, to avoid soiling the sheet (p. 57). Lettering and figuring should be well done as described ; scrawled cross-hatching must be avoided.

4. Have the scale or scales clearly indicated (p. 61).

5. Be numbered, dated and signed. Usually the number is placed in the top right-hand corner and the student's name with the date underneath is printed in the bottom right-hand corner. In drawing offices it is the usual practice to put the title and number at the bottom of drawings—this facilitates searching for drawings which are stored in drawers.

SKETCHING.—In addition to preparing drawings to scale, students are required to make freehand sketches of constructional details, etc. Many of the sketches made by the teacher on the blackboard are copied by the student. Sketches direct from constructional models are sometimes called for. Later in the course of his work he may be required to make sketches of building sites, portions of buildings in course of construction, etc. for the purpose of progress reports, existing buildings requiring alterations and repairs, and the like. He is also strongly recommended to make sketch details of work in course of erection and of completed buildings, a one metre rule (see 1, Fig. 155) being used for obtaining dimensions ; such sketches should be filed for future reference. Sketching should therefore be regarded as an essential accomplishment.

Sketches are made with the pencil and occasionally in ink. For the former a well-conditioned H.B. pencil of good quality is best. The note-book or pad (p. 56) should be of good surfaced *plain* paper. It is recommended that from the outset no mechanical aid, such as a scale, ruler, set square, lined or squared paper, etc. should be employed. When one or other of these aids is resorted to the mechanical appearance of such sketches is usually unsatisfactory, and, what is more important, such assistance, if continued, results in an imperfectly trained eye, hand and sense of proportion.

Sketches should be neat, well-proportioned and workmanlike. They

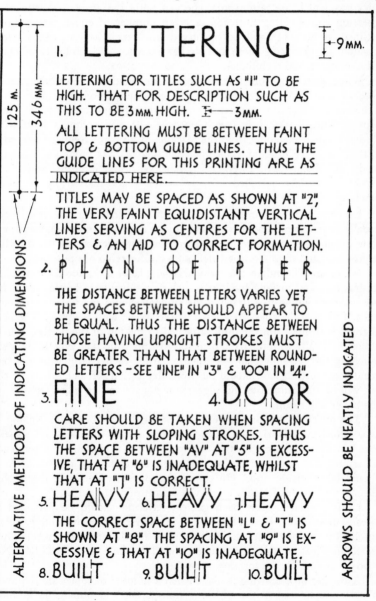

Fig. 31.

should be of a reasonably large size ; puny sketches are strongly condemned. Each freehand line should be drawn freely from end to end with a single stroke ; many students cultivate the bad habit of sketching each line by a series of short strokes. The beginner should practise simple exercises, such as sketching :—horizontal and vertical lines of varying thickness which are afterwards checked with the scale and squares ; bisecting and dividing lines into a given number of equal parts ; lines inclined at 30°, 45° and 60° ; curved lines such as arcs and circles. Assiduously practising these and similar exercises will gradually increase his confidence and technique.

When sketching from the blackboard and models, the student should observe with close concentration each detail and thoroughly understand it before commencing to sketch ; this will help the memory. Merely copying lines serves little useful purpose. It is a common fault with students, when sketching from the blackboard, to make frequent reference to the board—at least one glance per line sketched ; a preliminary careful study will obviate this. Any figuring must be *clearly* indicated and cross-hatching *carefully* sketched ; many sketches are marred by scrawled hatching.

INKING-IN, TRACING AND COLOURING.—A first year student is not usually required to ink-in, trace and colour drawings (although any spare time devoted to it will result in a general improvement in his drawing technique), and the following brief notes are for reference purposes.

Waterproofed Indian ink (which is jet black) is required for inking-in drawings and making tracings ; such is obtainable in bottles ready for use. Before inking-in a drawing the student should practise ruling inked lines of varying thickness (regulated by the screw) with the pen. The pen, charged with ink from the quill (attached to the top of the bottle), is held almost vertically with the fixed blade (as distinct from that which is hinged) touching the tee-square or set square. After each charging of the pen the outside of the blade should be wiped dry with a soft rag, otherwise the ink will be transmitted to the edge of the square and an inked smudge on the paper will result.

The pencilled drawing is cleaned down with a rubber before inking-in is commenced. Horizontal lines of the required thickness are usually inked-in first, followed by the verticals ; there is thereby less tendency for smudges to occur at the intersections. Hatching and dimension lines should be relatively thin. On completion, the pen should be thoroughly cleaned and the blade tension relieved by a turn of the screw.

Until proficiency has been gained, lettering should be pencilled as already described. This is then inked-in. A Heath " Telephone Pen " or similar broad-pointed nib is recommended for the style of printing illustrated in Figs. 29 and 31 ; a fine pointed pen-nib, such as Gillott No. " 303 ", is suitable for forming arrow heads, sketching, etc.

A tracing of a drawing is made when a number of copies are required. Tracing paper and cloth, which is transparent and obtainable in rolls, are used for this purpose ; the cloth, also known as tracing linen, is stronger than the paper and is capable of withstanding much handling without damage. The tracing paper or cloth is pinned and well stretched over the original drawing ; the longer it remains stretched before tracing is commenced the better. Tracing on paper is done in either ink or pencil, and in ink only on cloth. The latter has one side glazed and one unglazed. Tracing is usually done on the unglazed side and, to assist the ink to flow readily, the " chalk bag " (a piece of boiled linen containing powdered chalk) is applied ; if the chalk is well rubbed in, the cloth is stretched in the process and is therefore better conditioned.

Both paper and linen copies, called prints, are obtained from tracings in a printing machine. *Blue prints* are of paper with white lines on a blue ground ; *black-on-white prints* are of paper or linen and are white sheets with black lines. Briefly, in one type of machine, prints are produced in the following manner : The tracing is placed face downwards on a glass cylinder, a sheet of sensitized paper or cloth is stretched over the tracing (with the sensitive side next to the tracing) and a canvas cover is tightly stretched over them. An electric lamp is passed down the cylinder at the required speed ; the sensitized paper is moved, placed in a tray for development, washed, dried and its edges trimmed. The washing process causes these prints to shrink and hence the drawings on them are not true to scale. So-called *true-to-scale prints* can now be produced which, as implied, are not affected by shrinkage, but such are relatively expensive.

Working drawings are often coloured. Various portions of plans, the elevations and sections are coloured according to the materials of which they are composed. Whilst established office practice and individual preference may influence the colours used, the following are those commonly adopted for the materials stated : Brickwork, red produced by mixing crimson lake and vermilion ; masonry, burnt umber; concrete, Prussian blue ; carpentry or undressed timber, yellow ochre ; joinery or dressed woodwork, burnt sienna ; slating, Payne's grey or green obtained by an admixture of Prussian blue and yellow ochre ; roof tiling, as for brickwork but with an increase in the vermilion content. A denser colour is used for indicating materials in section than in plan or elevation.

Bottled inks in at least a dozen different colours are obtainable ; water colours are also sold in tubes, cakes and sticks. The colours are mixed in either saucers, palettes or tiles (divided into compartments). Sable brushes in various sizes are used for applying the colours.

The sheet or print to be coloured is stretched taut and pinned to the board, cleaned and preferably washed, *i.e.*, clean water is applied with a large brush (size No. 9 or 10) to each plan, elevation, etc. ; pre-watering assists the colour to flow freely and prevents a patchy appearance ; a clean double sheet of blotting paper should be applied immediately after each portion has been washed, and this should be also used to prevent the hand from touching and soiling the drawing during the colouring operation. Large light coloured washes are first applied, followed by the darker colours. At each dip of the brush the colour in the saucer should be well stirred and a full brush of the pigment collected. The colour should be kept flowing during the application. Large washes are commenced at the top and gradually worked from side to side with quick broad sweeps, finishing in one of the bottom corners ; the colour should be allowed to dry naturally, the blotter not being used for this purpose. The brush should be well washed in a glass of water between each change of colour. The tendency to lay on the colours too thickly must be avoided.

CHAPTER FOUR

FLOORS

Boarded and joisted floors : Single floors, sizes and spacing of joists, ventilation, trimming, strutting, boarding, joints ; double and triple floors ; double-boarding.

THE type of floor described in this chapter is the *suspended floor* consisting of structural or bearing timbers, called *joists* (or binders or girders), in addition to the *boards* which are used to cover them.[1]

CLASSIFICATION.—Boarded and joisted floors are usually classified into :—

(a) *Single Floors*.—This class has one set of joists called *common joists* or *bridging joists*.

(b) *Double Floors*.—Such a floor has one or more additional and larger joists, called *binders*, which wholly or partially support the common joists.

(c) *Triple or Framed Floors*.—As implied, there are three sets of joists in this class, *i.e.*, common joists which transmit the load to the binders, which latter are supported at intervals by larger joists called *girders*.

SINGLE FLOORS

This type of floor is generally adopted for relatively small buildings, such as houses, and the following description is of a typical ground floor and an upper floor of such a building.

GROUND FLOOR

The plan and two vertical sections of a ground floor of a portion of a building is shown in Fig. 32. A sketch of a portion of this floor is shown in Fig. 33. The plan (see p. 58) is that of a room which is 3·7 m wide. Three of its walls are outer or external walls and, as they are shown of solid brickwork, they must be at least 328 mm thick to prevent the penetration of rain to the inside surface and causing dampness ; an alternative type of external wall is shown in Fig. 36 and is known as a *cavity wall* (p. 76) ; if external walls are of stonework the minimum thickness is usually 400 mm ; external walls covered outside with suitable plaster can be reduced to 215 mm in thickness. The fourth wall shown on the plan is a division or partition wall and need only be 102·5 mm thick brickwork ; sometimes these division walls are 215 mm thick. Walls are constructed on foundations which are usually formed of concrete.

---

[1] Fire-resisting floors covered with boards, blocks, plywood, parquetry and cork are described in Chapter Two, " Joinery."

Concrete is a mixture of Portland cement (an artificial material made from chalk and clay, mixed in correct proportions, burnt and ground to a very fine powder), sand, and gravel or broken brick or broken stone. The width of the concrete foundation is twice the thickness of the wall to be supported, and the thickness of the concrete is at least 150 mm; the depth

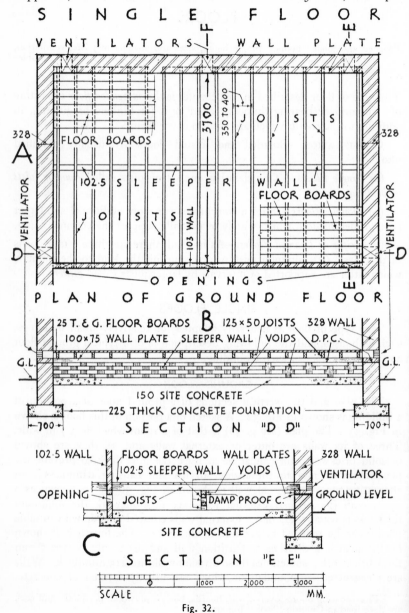

SINGLE FLOOR

VENTILATORS WALL PLATE

3100

350 TO 400

JOISTS

FLOOR BOARDS

328

A

VENTILATOR

102.5 SLEEPER WALL

FLOOR BOARDS

103 WALL

JOISTS

328

VENTILATOR

D D

OPENINGS

PLAN OF GROUND FLOOR

B

25 T. & G. FLOOR BOARDS 125×50 JOISTS 328 WALL

100×75 WALL PLATE SLEEPER WALL VOIDS D.P.C.

G.L. G.L.

150 SITE CONCRETE

225 THICK CONCRETE FOUNDATION

700 700

SECTION "DD"

102.5 WALL FLOOR BOARDS WALL PLATES 328 WALL

102.5 SLEEPER WALL VOIDS VENTILATOR

OPENING JOISTS DAMP PROOF C. GROUND LEVEL

SITE CONCRETE

C SECTION "EE"

SCALE 0 1000 2000 3000 MM.

Fig. 32.

of the foundations below the ground level varies, that of the external walls shown in Fig. 32 being 600 mm from the ground level to the top of the concrete (see B).

As shown, the joists are usually across the width (or smallest span—see below) of the room and are supported on the opposite walls. The amount of support at each end is 100 or 102·5 mm; this is known as the *bearing* or *wall-hold*; the joists in Fig. 32 have been given 102·5 mm wall-hold.

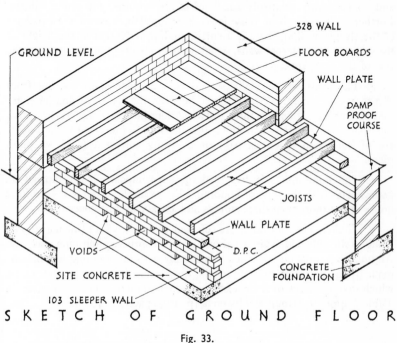

SKETCH OF GROUND FLOOR

Fig. 33.

The horizontal distance between the supports is called the *span*. The *clear span* is that between the internal faces of the walls supporting the joists; thus, in the example shown and assuming there was no intermediate support, the clear span would be 3·7 m. The *effective span* is the horizontal distance between the centre of each bearing, and in this example is 102·5 mm (2 × 51·25 mm) greater than the clear span.

According to general practice one or more *intermediate* supports are provided for ground floor joists when the clear span exceeds 1800 mm. These supports are called *sleeper walls* and are usually 102·5 mm thick (see p. 80). They are normally provided at a maximum distance apart of 1800 mm. In Fig. 32 one 102·5 mm thick sleeper wall is shown, and holes are formed during its construction for the reason stated on p. 80. These joists are thus supported at each end and in the middle; therefore, the clear span of these joists is (3700 ÷ 2) − 102·5 = 1747·5 mm.

Sizes of Joists.—The joists must be sufficiently large to safely support the load which the floor is required to carry. The sizes of joists depend upon (1) the span, (2) the distance between each joist, (3) the load on each square metre of floor and (4) the timber used. Regarding :—

1. The greater the span, the greater the load to be supported and therefore the greater the size of joists required. The provision of sleeper walls results in the clear span being reduced to a maximum of about 1·8 m and consequently relatively small joists are required for ground floors. Further, as the depth of these joists is reduced because of the existence of the sleeper wall(s), there is a corresponding reduction in the height of the external and internal walls, which in turn results in an economy in the cost of the brickwork.

2. The joists are normally spaced at a distance apart varying from 360 to 400 mm centres (the distance between the centre of one joist and that next to it). If 25 mm thick softwood floor boards (see p. 81) are used, this spacing is generally 400 mm ; this distance is increased if the boards are of hardwood (such as oak) or a strong softwood (such as pitch pine).

3. The minimum safe live load (or superimposed load) allowed on floors varies with the type of building. For example, it is 1·5 kN/m² for a house of not more than three storeys and 4·5 kN/m² for offices.

4. Timbers used for floors are referred to on pp. 38-43 (inclusive), *i.e.*, Douglas fir, redwood, whitewood or European spruce, pitch pine, oak, etc. All buildings must be constructed to comply with the Building Regulations made by the Secretary of State for the Environment. These include tables which give the sizes of floor joists and other timbers. The sizes given in Table I apply to Group 2 softwoods (p. 10), spaced at 400 mm.

TABLE 1

Maximum Clear Span in m	Size of Joist in mm (spaced at 400 mm Centres)	Maximum Clear Span in m	Size of Joist in mm (spaced at 400 mm Centres)
1	38 by 75	4·3	50 by 200
1·3	50 by 75	4·6	63 by 200 or
2	50 by 100		44 by 225
2·7	50 by 125	4·8	50 by 225 or
3·25	50 by 150		75 by 200
3·8	50 by 175	5·2	63 by 225
4	63 by 175		

The *approximate* depth in *millimetres* of 50 mm thick joists spaced at 400 mm centres may be found by dividing the span in *millimetres* by 20 and adding 20 to the quotient. Thus, the depth of a 50 mm thick joist with a span of 2400 mm would be $\frac{2400}{20} + 20 = 140$ mm.

Normally, the span for single floor joists of softwood is limited to 4·8 m.

WALL PLATES.—It is usual[1] to support the joists on wood members called *wall plates*. These are generally 100 mm by 75 mm and occasionally 100 mm by 50 mm. They (*a*) serve as suitable bearings (100 to 102·5) for the joists, (*b*) uniformly distribute loads from the joists to the walls below, (*c*) provide suitable means of bringing the upper edges of the joists to a horizontal plane to receive the floor boards and (*d*) afford a fixing for the ends of the joists.

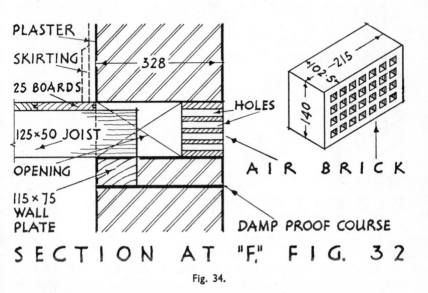

Fig. 34.

Wall plates are solidly bedded *level* on lime mortar by the bricklayer for the full length (or width) of the floor. A long straight-edge (p. 202) and spirit level (p. 219) are used to ensure that the wall plates are level (p. 84). On the score of economy it is a common practice to build the wall plates and ends of the joists into the walls as shown at A and C, Fig. 32, Figs. 33 and 34, and A, Fig. 36. Whilst this is the cheapest method it is not the best for, unless suitable precautions are taken, the timbers may become affected with dry rot (p. 31) ; well seasoned sound timber should be used, air spaces should be formed round the ends of the joists (see Figs. 33 and 34) and it is also advisable to apply a preservative (p. 26) to the wall plates and the ends of the joists.

A better form of bearing than the aforementioned is shown in Fig. 35. As indicated, two low 102·5 mm thick walls (sleeper walls) are built on the site concrete (p. 80) parallel to and about 50 mm from the main walls. The joists are secured to wall plates bedded on these sleeper walls and thus a free circulation of air round the timbers is provided.

---

[1] In the interests of economy wall plates are now generally omitted.

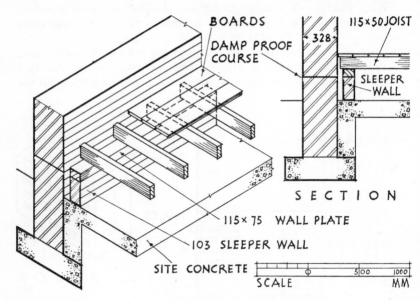

BOARDS

DAMP PROOF COURSE

115×50 JOIST

328

SLEEPER WALL

SECTION

115×75 WALL PLATE

103 SLEEPER WALL

SITE CONCRETE

SCALE

MM

S L E E P E R   W A L L   B E A R I N G

Fig. 35.

Another good support is that provided by the ledge or *offset* formed when the wall below the wall plate is increased 102·5 mm in thickness. Thus, the base of a 327·5 mm wall is increased to 430 mm and the wall plate is bedded on the 102·5 mm offset formed. This type of support is illustrated at B, Fig. 36 which shows the inner 102·5 mm portion of the cavity wall increased to 225 mm to form a 102·5 mm offset. Most external brick walls of houses (and other buildings) are now built of cavity construction in preference to solid construction. A cavity wall is usually 275 mm thick and consists of two 102·5 mm walls with a 70 mm cavity between. These 102·5 mm walls are tied together at intervals (900 mm horizontally and 450 mm vertically) with metal ties shaped as shown at R, Fig. 70 (see also C, Fig. 69). Two sections showing alternative construction at the base of a 275 mm cavity wall are given in Fig. 36 ; that shown at A is the most common, but the section at B shows sounder construction as a free circulation of air round the ends of the joists is assured. See also Figs. 48, 49, 50 and 58. Sometimes cavity walls are 387 mm thick with a cavity between the 102·5 mm outer leaf and the 215 mm inner portion (see Figs. 68 and C, Fig. 69).

Further reference to supports for wall plates is made on p. 101.

SUPPORTS AT CAVITY WALLS

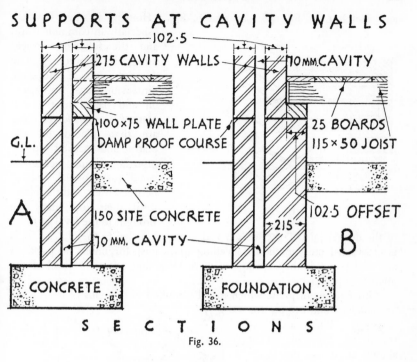

S E C T I O N S

Fig. 36.

Joints in long lengths of wall plates are formed as shown at G, Fig. 37. This is called a *longitudinal half lapped joint* or *scarf*. The vertical cut extends to half the thickness of each plate, and after the cut surfaces have been fitted together nails are driven in to make the joint secure. Occasionally intersections between wall plates are required; such are known as *right-angled half lapped joints* and are shown at H, Fig. 37; such joints are required at the corners of hipped roofs (p. 128) and as shown in Fig. 46.

When the ends of the joists rest upon the wall plates they are fixed by driving nails through their sides into the plates. The joists to be employed for a floor should preferably be of uniform depth. If they vary slightly in depth their upper edges are levelled by removing a portion of the wall plate as required to form a *housed joint*, as shown in the part section K

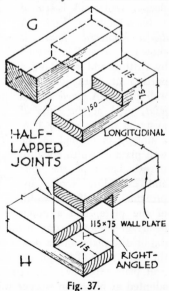

Fig. 37.

and sketch L in Fig. 38 (see also p. 84).

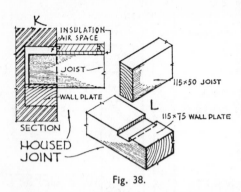

Fig. 38.

Housing is often omitted, especially in common work, and the following procedure is adopted. The deepest joists of the same dimension are supported on and nailed to the wall plates ; the shallower joists are then placed on the plates and packed up by inserting thin pieces of wood (or tapered wedges) between them and the plates until their top edges are brought to the level of those of the deeper joists ; the wedges and joists are nailed to the plates. This is a less satisfactory method of levelling up the joists for, unless the packings are well secured, they are apt to work loose and affect the level of the joists (see p. 84).

Other forms of joints which may be applied to the ends of deep joists are *notching* and *cogging* (or *caulking*). These are illustrated in Fig. 39. That shown at M has the lower edge of the joist cut to fit over the wall plate (such as may be supported by a sleeper wall) and is called a *single notched joint*. A *double notched joint* is shown at N, and is formed by cutting and removing a portion of both joist and wall plate. A *single cogged joint*, used at the ends of joists, is shown at O ; here a sinking is formed in the wall plate of width equal to the thickness of the joist and extending across approximately half the width of the plate, and a notch

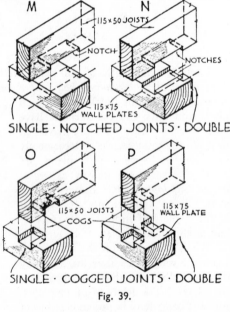

Fig. 39.

is cut on the lower edge of the joist to coincide with the uncut portion or *cog* on the plate. In a *double cogged joint*, shown at P, a notch is cut on the lower edge of the joist and this is fitted over the cog remaining on the wall plate after two sinkings have been formed ; such a joint may be adopted at plates on sleeper walls. Notching and cogging are not often used, although when joists are associated with binders (see p. 104) cogging may be resorted to (see B, Fig. 63 and H, Fig. 66).

At one time wall plates of mild steel or wrought iron were used instead of the more common timber type. Such flat bars were 50 to 100 mm wide by 3 to 10 mm thick (75 by 5 mm is a common size), of suitable length, and well painted before being bedded level on the wall. Such a wall plate is shown in Fig. 40.

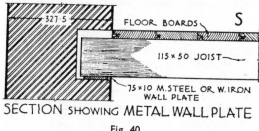

SECTION SHOWING **METAL WALL PLATE**

Fig. 40.

Wall plates are now generally omitted and the joists are supported directly on the brickwork. Such construction should only be adopted when the brickwork has been carefully built, *i.e.*, the brick course on which the joints are to rest should be at the correct height (the underside of the joists) and level throughout. When wall plates are omitted in inferior work the ends of the joists are packed up with pieces of slate, etc. in order to bring their top edges level ; this is an undesirable practice, as repeated vibration tends to disturb such bearings, resulting in unequal settlement of the joists and an uneven floor surface.

As already stated (p. 73) it is usual to fix joists across the shortest span. A space of about 50 mm should be left between the wall and the first joist which is parallel to it. When joists forming floors of adjacent rooms run in the same direction, the overlapping ends on the division walls are nailed to each other and to the wall plates (see Fig. 41).

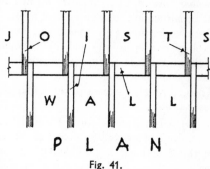

Fig. 41.

DAMP PROOF COURSES.—In order to prevent water absorbed from the adjacent ground rising up the walls and causing dampness in the walls and affecting timber members supported by them, a continuous layer of an impervious material, called a *horizontal damp proof course*, is provided. The absence of damp proof courses has been a frequent cause of dry rot (p. 32) in wood members such as wall plates and joists. There are several materials used to form damp proof courses. That most generally used is called *fibrous asphalt felt*. This consists of hessian (woven jute cloth) or felt which is impregnated with bitumen or asphalt (a dark-coloured impervious material). It is obtained in rolls 22 m long and in various widths from 102·5 to 915 mm, and bedded by the bricklayer (or mason) on mortar spread on the walls. It should be lapped 75 mm where longitudinal joints occur, and full width at crossings of walls and at corners. Another good damp proof course consists of two layers of slates embedded in cement mortar. The minimum height of the damp proof course above the ground level is 150 mm. Damp proof courses are shown in Figs. 32, 33, 34, 35, 46, etc.

VENTILATION.—Attention is drawn to the reference to dry rot which appears on pp. 31–33. It is there stated that one of the precautions which should be taken to prevent the onset of this virulent disease is the provision of an adequate circulation of fresh air round all timbers, and hence the need for (*a*) *through* ventilation under ground floors of timber construction, and (*b*) air spaces round the ends of joists. Regarding :—

(*a*) Through ventilation is provided by air bricks or cast iron ventilating grates which are built at intervals in the external walls, and by forming voids in sleeper and division walls during their construction. Air bricks are of various sizes ; a sketch of one form is shown in Fig 34 ; as indicated it is well perforated ; such air bricks should be built in a wall at about 1·8 m intervals. Air bricks or ventilators are shown in Figs. 32, 34, 44 and 46.

Typical means of ventilation under a ground floor are shown in Fig. 32. Three air bricks are shown in the top external wall and one is placed near the 102·5 mm division wall in each of the side outer walls. The sleeper wall is *honeycombed*, *i.e.*, voids are formed by simply omitting bricks during its construction ; two alternative forms of voids are shown at B, Fig. 32. See also sketches, Figs. 33 and 46. Damp proof courses must be provided to sleeper walls, as indicated. Three openings are formed in the 102·5 mm division wall (see A, Fig. 32) where it is assumed that the adjoining room has a floor of similar construction [1] with air bricks in its external walls. Thus, through currents of air under the whole of the ground floor are assured.

(*b*) Air spaces are formed round the sides and tops of built-in joists as shown in Figs. 33, 34, 38 and 40. Free circulation of air round the ends of the floor joists is provided when the joists are supported on sleeper walls, as shown in Fig. 35, and when supported on offsets (B, Fig. 36).

SITE CONCRETE.—The area of a building below wood floors must be covered with an impervious material in order to exclude dampness. The material usually adopted for this purpose is concrete (p. 72) and the Building Regulations stipulate that this must be 100 mm thick. This is known as *site concrete* or *surface concrete*, and is shown in Figs. 32, 33, 36,

---

[1] If this adjacent floor is of concrete construction, three 100 mm dia. drain pipes would be embedded in the concrete, *i.e.*, one continuous from each of the three holes in the division wall to an air brick in the bottom outer wall corresponding to that at the top of plan A, Fig. 32.

44, 46 and 48. Besides excluding dampness, site concrete prevents the growth of vegetable matter and the admission of ground air.

The site concrete should be well brushed, and all debris below the floor removed, before the floor boards are fixed. Dry rot may be caused by small pieces of wood and shavings left below a floor (p. 32) becoming affected, probably on account of dampness, and spreading to the members of the floor.

After the joists have been levelled, with their upper edges in the same plane (see p. 84), they are now ready to receive the floor boards.

FLOOR BOARDS.[1]—Softwoods used for floor boards include Douglas fir, pitch pine, red pine, redwood, sitka spruce, western hemlock and whitewood. The hardwoods employed for this purpose include beech, white elm, gurjun, iroko, oak, rock maple and teak. Pitch pine and the hardwoods are used for first-class floors ; Douglas fir and redwood are often specified for ordinary good class work ; whitewood is employed for cheaper work.

The *stock* widths of *softwood* boards are 115, 125, 150, and 175 mm ; the thickness varies from 19 to 32 mm. The narrower the boards the better, for then the shrinkage of each will be reduced to a minimum, the joints will not appreciably open, and there will be less tendency for the boards to cup (p. 34). Hence softwood boards in " narrow widths " (about 90 mm wide) are specified for good class work. For average work, 25 mm thick boards are used when the joists do not exceed 400 mm centres ; 32 mm thick boards are used for better class work. *Hardwood* boards are obtainable from stock in 65, 75, 100 and 112 mm widths, and the thickness of these varies from 19 to 32 mm. As mentioned on p. 18, the boards should be rift or quarter sawn for first-class work, and if tangentially sawn (p. 17) boards are used they should be nailed with the heart side downwards (p. 19).

The above sizes are known as *nominal* or *stuff* sizes, *i.e.*, those after the boards have left the saw. After the boards have been shaped as required (see below) and *dressed* or *wrought* or *planed* (see p. 209) the sizes are reduced and are known as *net* or *finished* sizes. Thus, a floor board has one side (which is laid uppermost) and both edges planed, and a 150 mm by 25 mm (nominal size) board is reduced to about 144 mm by 22 mm net, and a 32 mm (nominal) board has a finished thickness of 29 mm ; the net width includes the tongue (see Q, R, V and W, Fig. 42). Boards are usually obtained up to 5 or 6 m lengths, although longer boards are obtainable.

JOINTS.—The various *edge* or longitudinal joints between floor boards are (a) tongued and grooved, (b) square, (c) rebated, (d) rebated, tongued and grooved, (e) splayed, rebated, tongued and grooved and (f) ploughed and tongued. These are illustrated in Fig. 42.

(a) *Tongued and Grooved or Feathered and Grooved Joint* (see R and U).— This joint, abbreviated to " t. and g." or " f. and g.", is used more

---

[1] Flooring is included in joinery and is mentioned here in order to make the description complete.

frequently than any other for good work. A narrow projecting tongue or feather is formed just below the middle along one edge and a groove is cut along the other. The tongue is slightly smaller than the groove ; thus, for a 10 mm wide tongue the groove is approximately 12 mm deep. The tongues are sometimes slightly rounded off so as to facilitate the laying of the boards and prevent them being damaged during the process.

The *matched joint* somewhat resembles the t. and g. joint, except that both sides of the tongue are splayed (like the splayed surface at w) and the groove is formed to correspond.

(b) *Square, Straight or Plain Joint* (see P).—The edges are cut and planed at right angles to the face, when they are said to be *shot*, *butt jointed* or *straight-edged*. This joint is never used for good class flooring unless the boards are to be covered by another layer of boards to form what is called a *double boarded floor* (p. 111). The chief defects of a floor covered with butt jointed boards are its unsightly appearance when the boards shrink, draughts which are caused (especially if it is a ground floor) by the admission of air through the joints, and dust which is blown through open joints. It is commonly adopted for roof boarding (pp. 115 and 140).

(c) *Rebated Joint* (see Q).—A 10 mm wide tongue, one-third the thickness of the board, is formed along the lower edge and fitted into a slightly wider *rebate* or *rabbet* (rectangular groove) on that adjacent. This lapped joint is very rarely used for edge joints, but is sometimes adopted in good work for heading joints (see below).

(d) *Rebated, Tongued and Grooved Joint* (see v).—This is an excellent joint which is sometimes employed for hardwood floors when *secret nailing* is adopted, *i.e.*, nails are driven through the tongues into the joists and are therefore concealed (see nail indicated by broken lines). Hence the appearance is very satisfactory and for this reason secret nailing is preferred to *top nailing* (see p. 86).

(e) *Splayed, Rebated, Tongued and Grooved Joint* (see w).—As shown, this is another secretly nailed joint, used in hardwood flooring, and is an improvement upon that at v because of the thicker and stronger tongue.

(f) *Ploughed and Tongued Joint* (see x).—Grooves are formed or " ploughed " (as the plough plane—see 2, Fig. 158—was used for making the grooves) in the square edges of the boards to receive hardwood tongues or " slip feathers ". It is now rarely employed unless very thick boards are required and where the ordinary tongued and grooved joints (R) would result in an excessive waste of material in forming the tongues.

Heading or End Joints.—Wherever possible, the boards should be long enough to reach from wall to wall of a room in order to avoid end or heading joints. Where such joints are necessary, as for large floors, they occur at the joists and usually take the form of the square joint shown at P, Fig. 42. Each adjacent board is cut to cover half the thickness of the joist below, the ends are closely butted together, and four nails are driven in, two on each side of the joint, as indicated at A, Fig. 58.

Rebated heading joints (Q, Fig. 42) are sometimes specified for good work, the top board only at each joint being twice nailed. Another good joint, provided it is accurately made, is the *splayed* or *bevelled heading joint* (see Y, Fig. 42) ; the ends are splayed to give a tight fit, and two nails (as shown B, Fig. 58) are hammered in at an angle (as indicated at Y, Fig. 42). Occasionally heading joints are tongued and grooved (R, Fig. 42) as for edge joints.

FLOOR BOARD JOINTS

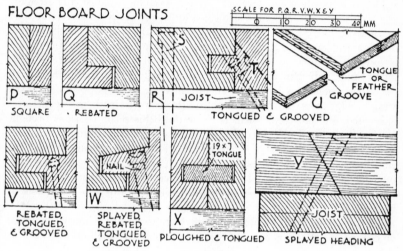

Fig. 42.

The appearance of the work is spoilt if the heading joints form one continuous line over the same joist, and unevenness occurs when the boards shrink. They should be laid to *break joint* (the heading joints being on different joists) as shown in Fig. 58. Sometimes they are arranged with not more than three continuous heading joints, but the appearance is not so satisfactory.

MANUFACTURE OF FLOOR BOARDS.—The methods of converting logs into floor boards are described on p. 17. Labours such as tonguing, grooving, rebating and planing are carried out in one operation by a machine called a *planing and matching machine*. This is a powerful machine which can very rapidly convert sawn boards into tongued and grooved, etc. boarding. Thus, the sawn boards, which are often automatically fed into the machine, are first planed to a smooth finish on the lower face as they proceed horizontally, followed in turn by the planing and grooving on one edge, tonguing on the other edge and reducing to the correct width, and finally reducing to the required thickness by cutting the top face just before leaving the machine. From 60 to 156 m of tongued and grooved boarding can be produced per minute by this class of machine.[1]

[1] This machine is described in Chapter One, " Joinery."

As stated on p. 89, *hardwood mitred margins* or *borders* are sometimes fixed round fireplace hearths.

LEVELLING JOISTS.—The top edges of the joists must be level throughout. A spirit level (3, Fig. 162) and long straight-edge (p. 202) are used " to test for level." Assuming that the joists are to bridge the shortest span of a room and, commencing at one of the end walls, the wall joist is placed in position and levelled. When levelling the joist, the straight-edge is placed on its top edge near the middle, and the spirit level is applied to the straight-edge, as shown at 7, Fig. 155. If the joist is level, the bubble of the spirit level will be in the centre of its run ; if not, the higher end is lowered by housing (p. 78) it into the wall plate as required. The joist is then nailed to the wall plate. The wall joist at the opposite wall (or, if the room is a long one, an intermediate joist) is positioned ; the straight-edge is rested on both wall joists near one end, the spirit level is applied, and any adjustment is made until this second joist is brought level with the first. The second wall joist is then levelled (in the direction of its length) like the first. A check is provided if the opposite ends of the wall joists are tested for level. After nailing this second wall joist, the intermediate joists are fixed after all of their top edges have been made to conform to the straight-edge when applied to the two wall joists at both ends.

If the joists vary in depth, and if they are to be either housed, notched or cogged (p. 78), it is customary to use two of the shallower joists near the end walls and make the deeper joists level through with these by notching, as required, the wall plates. If, however, housing, etc. is omitted, it is usual to start with two of the deepest joists and pack up the shallower intermediate joists until their top edges are level with the straight-edge (see p. 78). The joists are now ready to receive the floor boards.

LAYING FLOOR BOARDS.—The longitudinal joints must be as close as possible before the boards are nailed. Three methods of effecting this are by (1) cramping, (2) folding and (3) dogging. They are illustrated in Fig. 43.

1. *Cramping.*—This method is chiefly adopted, it being expeditious and effective in ensuring tight joints. One or preferably two metal appliances, called *cramps*, are required. The plan of one type of cramp is shown at A. The following is the procedure when laying the boards when they are tongued and grooved and are to be top nailed (see p. 86) : Starting at one of the walls, a board is placed at right angles on the joists at about 16 mm (approximate thickness of the plaster) away from the wall and nailed to the joists. Three to six boards are laid loosely upon the joists with their tongues inserted in the grooves. Two cramps are placed temporarily over joists which are some 0·6 to 1·8 m from the ends of the boards. Each cramp is fixed to the joist as shown at A by rotating the arm D in the direction of the arrow " 1 " ; this causes E to rotate towards

the joist in the direction of arrow " 2 " when the grooved surface on E and the sharp metal points at F (which project from the side and under the top plate G) cause the cramp to grip the sides of the joist. A rough strip of wood is inserted between the last laid board and the plate c to protect the edge of the board, the arm B is rotated in the direction of arrow " 3 ", and this causes the plate c to move forward in the direction of arrow H to exert considerable pressure on the boards until the joints between them are completely closed. Both cramps are operated together. The boards are then nailed as described on p. 86, the cramps and the protecting strip are removed, and the operation is repeated on the next set of boards. As the work proceeds towards the opposite wall, the last few lengths of boards cannot be cramped owing to lack of space. These boards may be brought tight by using a short piece of floor board which is inclined with the upper edge against the wall and the lower edge against

METHODS OF LAYING FLOOR BOARDS, ETC.

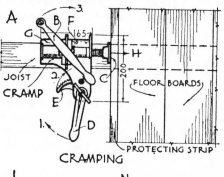

CRAMPING

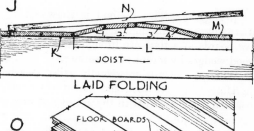

LAID FOLDING

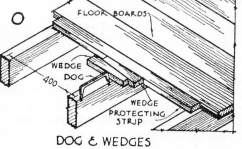

DOG & WEDGES

Fig. 43.

the protecting strip ; a few sharp knocks with a heavy hammer on the top end of the inclined board will tighten the joints preparatory to nailing. The last board may have to be sawn longitudinally and fitted to the wall (or about 16 mm from it).

It is usual to lay the boards with the grooved edge of the last board next to the protecting strip, otherwise the strip should be grooved to just clear the tongue to prevent damage to the latter when clamp pressure is applied. This method can be adopted when the heading joints are " breaking joint " (p. 83).

2. *Folding*.—When cramps are not available, the joints between the boards may be closed by " jumping them in " or " laid folding ". This method is shown at J, Fig. 43. Assuming that the floor has been laid up to K, a floor board M is securely nailed at a distance L, which equals the width of the five boards when placed in position tightly by hand less 6 to 12 mm, depending upon the width of the boards. The four boards " 1 ", " 2 ", " 3 " and " 4 " are then placed as shown (with the tongues engaging in the grooves) and forced into position by jumping on board N laid across them. If the joiner has an assistant, two boards are used, one near each end, to spring the boards into position.

In this method the first board is well fixed next to a wall, as in cramping, and the first set or *bay* of boards is laid folding against it. The boards in each bay are of the same length, and the heading joints are therefore continuous for at least four boards ; the bays, however, should break joint as the work proceeds in order to avoid continuous heading joints from wall to wall (p. 83). The appearance of the floor is not so satisfactory as that of a clamped broken jointed floor and the joints are usually not so close.

3. *Dogging*.—This method is shown at O, Fig. 43, and is another alternative to cramping. A metal *dog* is driven into a joist, and four or five boards at a time are brought close together by tightening the hardwood wedges by means of a hammer.

When the boards are secretly nailed (see below), and each board has therefore to be cramped and nailed separately, it is a common practice to cramp each board with the aid of a strong chisel which is driven into the top of a joist close to the protecting strip and used as a lever. The blade of the chisel is forced against the strip until the pressure closes the joint.

NAILING BOARDS.—Floor boards are generally secured by oval nails (see A, Fig. 152), the length of which should be twice the thickness of the boards. They are either *top nailed* or *secret nailed*.

When top nailed, two nails are driven through each board to every joist which it covers, including two nails at the ends. The nails are about 25 mm from the edges, and after the boards have been fixed, the heads of the nails are driven below the surface by using a hammer and punch (see I, Fig. 160). Tongued and grooved boards are usually top nailed as shown by broken lines at S, Fig. 42 ; rebated, and ploughed and tongued boards are also top nailed. In better class work, such as hardwood flooring, t. and g. boards are secretly nailed as shown in the two positions at T, Fig. 42, the higher position being the better of the two as the tongue is less likely to be damaged ; the heads of the nails are also punched. As stated on p. 82, hardwood boards which are rebated, tongued and grooved jointed or splayed, rebated, tongued and grooved jointed, are secretly nailed ; the oval wire nails are driven through the tongues and their heads are punched (see V and W, Fig. 42).

Water and gas pipes and electric cables are frequently run below the floor boards. It is necessary that such pipes and cables be readily accessible, and therefore the boards over them should be screwed (see the flat-headed screw at K, Fig. 152), and not nailed, to permit of their easy removal.

CLEANING OFF AND PROTECTING FLOORS.—Wood floors on completion should be *traversed* or *flogged* and cleaned off. This consists of planing the boards to a level and smooth surface either by hand or machine. In cheap or unimportant work any irregularities are simply removed by the application of the *jack plane* (see 1, Fig. 158). For better class work, especially if the flooring is of hardwood, greater care is exercised and more labour expended to ensure a smooth level surface. Thus, if done by hand, the boards are first traversed with a jack plane in a diagonal direction across the floor ; any irregularities are then removed by applying the *trying plane* (see 6, Fig. 158) diagonally until the surface is level throughout, *i.e.*, when a long *straight-edge* (see p. 202) is applied in any position on the floor it should contact the surface throughout its length— there should be no " hollows " ; finally, the floor is finished off by the *smoothing plane* (see 7, Fig. 158) until a smooth surface has been obtained. This hand dressing has been largely superseded by the machine, especially on large floors, an electrically driven portable *machine planer* being used to plane the surface in a fraction of the time occupied by hand labour. When hardwood floors, after being planed, are required to be wax polished they are first scraped (see p. 217 and 7, Fig. 162), rubbed smooth with glass-paper (pp. 217, 220) and finally oiled or waxed and polished : polishing can be done more expeditiously by a *machine sander*.

Immediately on completion, floors should be protected against damage during subsequent building operations (such as plastering) by covering them with about 50 mm thickness of sawdust (waste product from the workshop in which sawing and planing machines, etc. are employed). This prevents plaster (which may cause much damage when trodden with the feet), paint and dirt from soiling and scratching the boards ; sawdust also absorbs moisture from plaster droppings, etc. Care should be taken to ensure that the floors remain completely protected until the whole of the work has been finished, any portions which may have become exposed being at once recovered.

It is essential to defer floor-laying until the building has " dried out " or is as free from moisture as possible.

TRIMMING.—When a fireplace is provided in a ground floor room certain adjustments in the construction of the floor are necessary to accommodate the front hearth. The projecting brickwork (or stonework) of a fireplace is called the *chimney breast*, and that at each side of the *fireplace opening* (in which the kitchen range, etc. is fixed) is called a *jamb*. The *back hearth* is that within the jambs, and the *front hearth* is

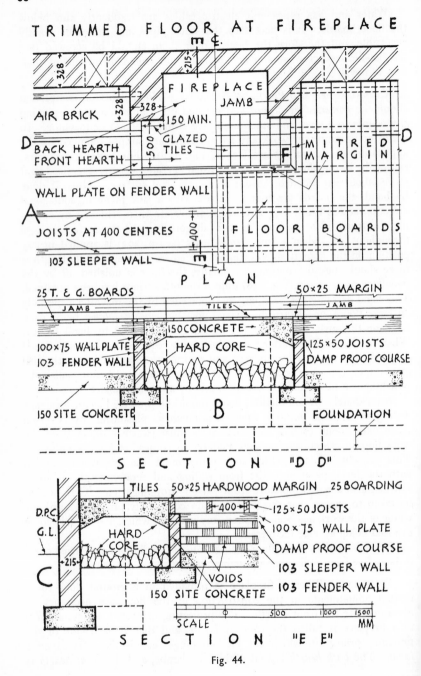

TRIMMED FLOOR AT FIREPLACE

PLAN

- AIR BRICK
- BACK HEARTH
- FRONT HEARTH
- WALL PLATE ON FENDER WALL
- JOISTS AT 400 CENTRES
- 103 SLEEPER WALL
- FIREPLACE JAMB
- 150 MIN.
- GLAZED TILES
- MITRED MARGIN
- FLOOR BOARDS

SECTION "D D"

- 25 T. & G. BOARDS
- 50×25 MARGIN
- JAMB
- TILES
- JAMB
- 150 CONCRETE
- 100×75 WALL PLATE
- HARD CORE
- 125×50 JOISTS
- 103 FENDER WALL
- DAMP PROOF COURSE
- 150 SITE CONCRETE
- FOUNDATION

SECTION "E E"

- TILES
- 50×25 HARDWOOD MARGIN
- 25 BOARDING
- D.P.C.
- G.L.
- HARD CORE
- 125×50 JOISTS
- 100×75 WALL PLATE
- DAMP PROOF COURSE
- 103 SLEEPER WALL
- 103 FENDER WALL
- VOIDS
- 150 SITE CONCRETE

SCALE   500   1000   1500   MM

Fig. 44.

in front of the chimney breast. Ground floor fireplaces are shown in Figs. 44, 46 and 48. Hearths must be of concrete or other non-combustible material. The Building Regulations require that a hearth must be at least 125 mm thick (see B, Fig. 44), it must extend at least 150 mm beyond the opening at each end (see A, Fig. 44) and its minimum projection from the breast is 500 mm (see A, Fig. 48 and A, Fig. 44). These drawings show the necessary adjustment or trimming of the floor which is required. A wall is built round the fireplace to retain the concrete hearth and support a portion of the floor. This is called a *fender wall* and its thickness may be 102·5 mm (as in Fig. 44) or 215 mm (see Fig. 48), depending upon the height and the load which it has to support.

The front hearth shown in Fig. 44 consists of 150 mm concrete on which glazed tiles are bedded on mortar. It is supported on broken bricks or stone, called *hard core* (or *penning* or *filling*), which should be well packed. In inferior work it is a common practice to dispense with hard core and fill in with earth which is rammed ; this is very unsatisfactory, as moisture (especially if the site is damp) may be transmitted from the filling to the wall plates and ends of the joists, and may thereby cause dry rot (pp. 31-33).

The arrangement of the joists should be noted. In good work it is customary to fix a *hardwood margin* round a front hearth. This ensures a more accurate finish and a neater appearance than are presented if the

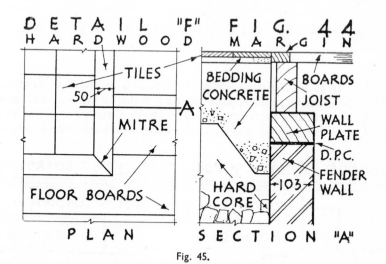

Fig. 45.

floor boards are stopped against the concrete or tiles. The margin shown in Fig. 44 is the same thickness as the floor boards and it is mitred at its two external angles. The enlarged detail in Fig. 45 shows the construction more clearly. Note that the fender wall is provided with a damp proof course.

Fig. 46 is an isometric sketch of the fireplace and floor detailed in Fig. 44.

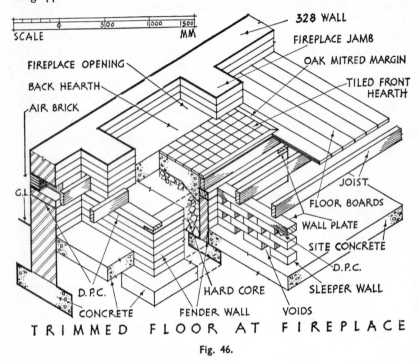

SCALE   MM

FIREPLACE OPENING
BACK HEARTH
AIR BRICK
G.L.
D.P.C.
CONCRETE
FENDER WALL
HARD CORE
VOIDS
SLEEPER WALL
D.P.C.
SITE CONCRETE
WALL PLATE
FLOOR BOARDS
JOIST
TILED FRONT HEARTH
OAK MITRED MARGIN
FIREPLACE JAMB
328 WALL

TRIMMED FLOOR AT FIREPLACE

Fig. 46.

Fig. 48 shows the trimming of the floor at a ground floor fireplace when the joists run in the opposite direction to that in Fig. 44. This also shows an alternative treatment of the front hearth whereby a 75 or 100 mm thick stone slab is provided as a support in lieu of the hard core. Enlarged

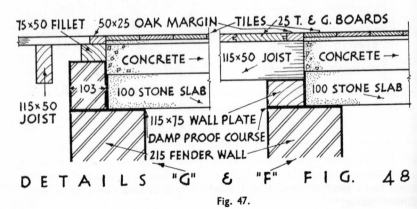

75×50 FILLET   50×25 OAK MARGIN   TILES   25 T. & G. BOARDS
CONCRETE →
115×50 JOIST
CONCRETE →
103
100 STONE SLAB
100 STONE SLAB
115×50 JOIST
115×75 WALL PLATE
DAMP PROOF COURSE
215 FENDER WALL

DETAILS "G" & "F" FIG. 48

Fig. 47.

vertical sections are given in Fig. 47; a 75 mm by 50 mm wood fillet (bedded on the wall) shown in detail G provides a fixing for the margin and the ends of the floor boards.

TRIMMED FLOOR AT FIREPLACE

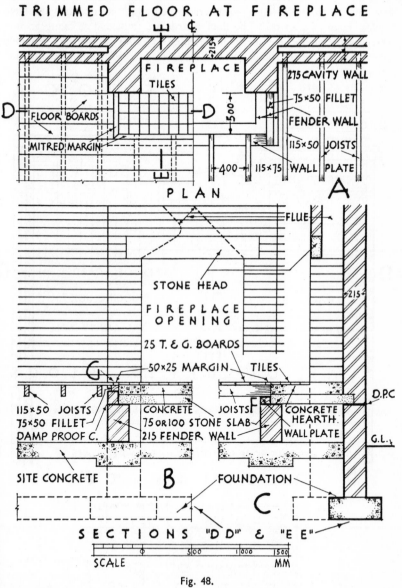

Fig. 48.

Ground floor joists are sometimes trimmed round fireplaces as described for upper floors (see p. 93). This is in lieu of the fender wall construction.

UPPER FLOOR

The part plan and two sections of an upper single floor are shown in Fig. 49. The external walls are of cavity construction (p. 76). The 225 mm by 50 mm bridging joists are spaced at 400 mm centres and have a clear span of 4·27 m. The floor boards are indicated as being 30 mm thick, although for average bedroom floors 25 mm boarding is generally specified. Details of the floor when the joists run in the opposite direction are given in Fig. 50.

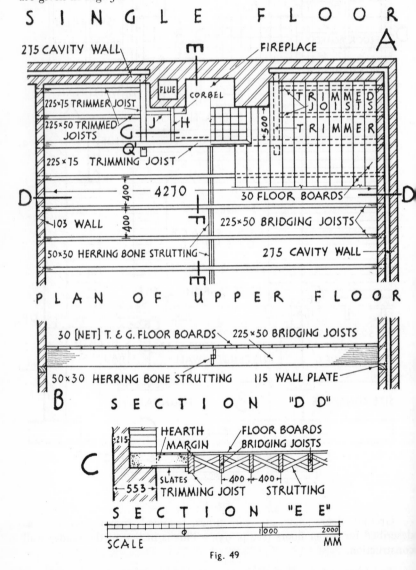

Fig. 49

TRIMMING.—Where fireplaces and openings (such as are required for stairs) occur some of the bridging joists cannot be supported by walls at both ends, and therefore additional wood members are necessary to receive the ends of the joists which have to be cut or *trimmed*. The trimmed opening at the fireplace shown in Fig. 49 has a thick joist, called a *trimming joist*, which is 500 mm (the width of the front hearth) from the fireplace and spans the full width of the room. The trimming joist partly supports

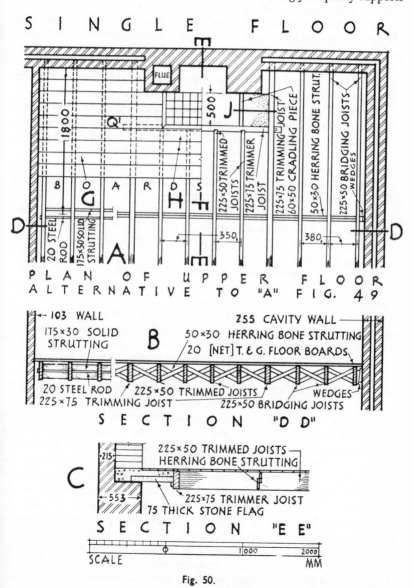

Fig. 50.

two short cross joists called *trimmer joists* or *trimmers*, and each trimmer
partly supports two *trimmed* or *tail* joists. In the alternative plan A,
Fig. 50, the two trimming joists support a trimmer and the latter carries
four trimmed joists. Thus, a trimming joist is one which has one or more
trimmers connected to it, and a trimmer receives the cut bridging joists
called trimmed joists. The arrangement of the joists in these two plans
is in accordance with the Building Regulations controlling the construction
of wood floors adjacent to fireplaces.

Trimmer and trimming joists should be thicker than bridging joists on
account of the greater weight which they have to support. Tables are
provided in the Building Regulations which give the sizes of beams which
support other joists. It is usual to make the thickness of a trimmer and a
trimming joist 25 mm greater than that of the bridging joists. Thus in
Figs. 49 and 50 the trimmers and trimming joists are 75 mm thick.

JOINTS.—The joints between trimmers and trimming joists, and
between trimmed joists and trimmers, should be as strong as possible.
When making these joints only the minimum amount of wood should
be removed, otherwise the joists will be unduly weakened.

The following joints are adopted between joists at trimmed openings
in floors : Tusk tenon joint, dovetailed housed joint, bevelled housed
joint and square housed joint.

Tusk Tenon Joint.—This is the strongest form of joint used in floor
construction, and for this reason it should be employed for the connection
between the trimmer and trimming joists, as at Q¹ indicated on the plans
in Figs. 49 and 50.

A *tenon* is a projecting piece formed at the end of a timber member ;
it is inserted in a corresponding hole or *mortise* (or *mortice*) which has
been cut in an adjoining wood member.

A plan, section, end view and isometric sketch of a tusk tenon joint
are shown in Fig. 51. The tenon which is cut at the end of the trimmer
is shown in the *centre* of the latter ; it should be sufficiently long to pass
through the mortise formed in the trimming joist and project some 100 to
125 mm beyond. Sometimes the *underside* of the tenon is made to coincide
with the centre line of the trimmer ; although this forms a somewhat
stronger joint, it is more difficult to make the joint tight. The projecting
piece or *tusk*, formed under the tenon, transmits most of the weight from the
trimmer to the trimming joist and enters the trimming joist for ⅙ to ¼
the thickness of the latter. The bevelled or slanting portion above the
tenon, called the *horn* or *haunch*, strengthen the tenon. The trimmer is
brought tight against the trimming joist by driving a wood wedge down
through a hole formed in the tenon ; the side of the hole (shown by a
thick line in the section) should be cut to the same angle as that of
the tapered wedge, and this hole must be long enough to allow the trimmer
to be forced in the direction of the arrow until the joint is tight.

A modified form of tusk tenon joint, called a *bevelled haunched joint*,
is sometimes adopted between a trimmer and each of the trimmed joints
where it is not possible to have projecting tenons on account of the hearth.

This is shown by broken lines at A, Fig. 55. When the tenon formed on the trimmed joint has been inserted, each side of the mortise in the trimmer is slightly pared to receive a wedge which is driven in to tighten the tenon;

150 mm wire nails are then hammered in from the top and sides of the trimmer and through the tenon.

A further modification consists of a shorter tenon (with tusk) which enters a corresponding mortise in the trimmer. Long nails driven in from the top of this joint make the joint secure.

Dovetailed Housed or Notched Joint (see Fig. 52).— This is another good joint which is used to connect trimmed joists to a trimmer. The end of the trimmed joist is formed to correspond to the

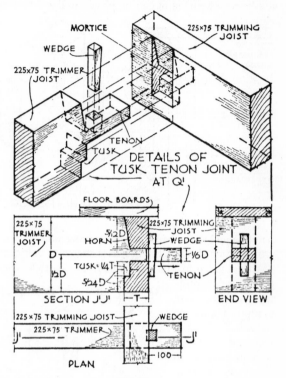

Fig. 51.

housing (one edge of which is dovetailed as shown) made in the trimmer to receive it, and is dropped into the housing. Long nails are then driven in slant-wise from the outer face of the trimmer and through the end of the trimmed joist. This joint is indicated at the trimmer shown at A, Fig. 50 (see also the broken lines at the trimmer joist in Fig. 56).

Bevelled Housed Joint (see N, Fig. 53). — This is a cheaper but not an effective alternative to the dovetailed housed joint and is used for the same purpose. It is known as a *half-depth joint*, as the depth of the housing equals half the depth of the joist. The members must be nailed securely. This joint is shown by broken lines at B, Fig. 55 (see p. 97).

Square Housed Joint (see O, Fig. 53).—This is another half-depth joint which may be adopted at the connections between short trimmed joists (such as those at A, Fig. 49) and a trimmer joist.

HEARTHS.—It has been stated on p. 89 that a front hearth shall be at least 500 mm wide, it must have a minimum thickness of 125 mm, and it shall extend at least 150 mm beyond each side of the fireplace opening. A further Building Regulation in regard to hearths is that combustible material (other than timber fillets supporting the edges of a hearth where it adjoins a floor) is not to be placed under a fireplace hearth nearer than 250 mm vertically below the top of the hearth, unless such material is separated from the underside of the hearth by an air space of at least 50 mm.

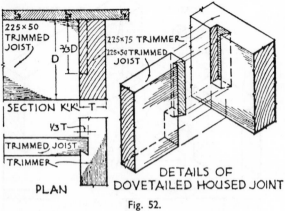

PLAN

DETAILS OF DOVETAILED HOUSED JOINT

Fig. 52.

The plan at A, Fig. 49, shows the front hearth, covered with glazed tiles, having the required projection and a length equal to the width of the opening plus 150 mm on either side. A cross section through the hearth is shown at C, Fig. 49, and an enlargement of this is given in Fig. 54. The front hearth of concrete is formed *in situ* (or permanent position) and therefore a support must be provided for the concrete. This support is shown in Fig. 54 to consist of slates which rest on the fireplace opening and on corbels at either side of the opening; at the front the slates are carried on a 50 mm by 40 mm fillet or bearer nailed to the trimming joist.

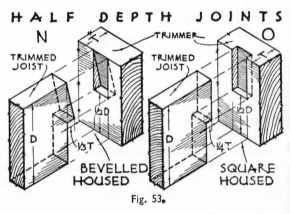

Fig. 53.

A short joist is provided to afford a support for the boards at each end of the hearth, and between the fireplace jamb and the trimming joist (H on the plan in Fig. 49). This joist H is called a *cradling piece*; it is housed at one end into the trimming joist, and the other end rests upon a short brick corbel (or projecting brick) as it must not enter the wall owing to the proximity of the flue from the ground floor fireplace. These details are shown in Fig. 55. Note: (1) the brick corbel which supports one end of the cradling piece, (2) the bevelled housing B (indicated by broken lines) of piece J, (3) the bevelled haunched joint A (broken lines) between the trimmed joist and trimmer, and (4) the hardwood mitred margin (see p. 89) which is also shown in Fig. 54.

S E C T I O N   "E F"   F I G.   4 9

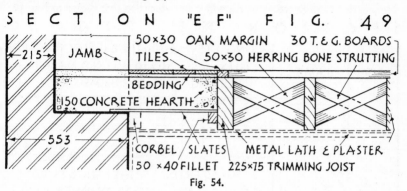

Fig. 54.

The ceiling below the floor is formed by metallic lathing which is nailed to the joists and bent down and nailed to the brickwork joints below where the hearth meets the wall. Two or three coats of plaster are then applied to form a 16 mm thick ceiling (see broken lines in Figs. 54 and 55).

An alternative form of front hearth is shown at A and C, Fig. 50 and Figs. 56 and 57. This hearth extends beyond the chimney breast to the trimming joists (see A, Fig. 50 and Fig. 57). A 75 mm stone or concrete slab is shown supported on the brickwork (or on a brick corbel course) along one edge, and upon wood fillets which are well nailed to the trimming joists and trimmer. Concrete is placed upon this stone to bring the thickness up to that required by the Regulations, and this is generally covered with the tiles. Concrete is used to form the back hearth which is brought up to the level of the front hearth. Provision must be made for securing the ends of the floor boards and the returned portions of the mitred margin. As shown (see J, Fig. 50 and Fig. 57) this takes the form of a 65 mm by 50 mm cradling piece or fillet which is partly embedded in the concrete; this is splayed on one or both edges to give a firm hold in the concrete. The trimmed joists are shown dovetailed housed to the trimmer (see broken

S E C T I O N   "G"   F I G.   4 9

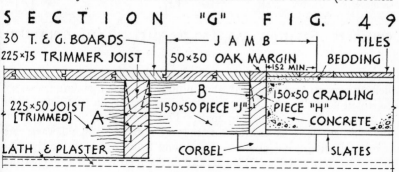

Fig. 55.

lines in Fig. 56.) Fires to buildings have been caused because the front hearths have not extended sufficiently beyond the fireplace openings. Hence, this front hearth construction is sounder than that shown in Fig. 49,

SECTION "E F" FIG. 50

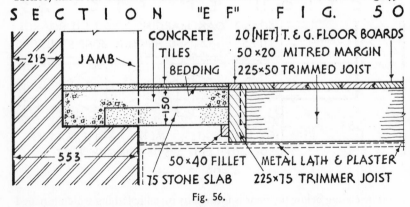

Fig. 56.

especially if the tiled portion, together with the mitred margin, is extended to the full width of the chimney breast.

An isometric sketch showing the trimming of the fireplace in Fig. 50 is given in Fig. 58.

STRUTTING.—With certain exceptions (such as floors of dance halls) floors should be as rigid as possible, otherwise undue stress may be trans-

SECTION "J" FIG. 50

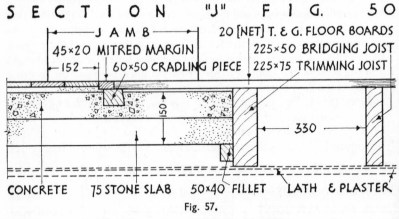

Fig. 57.

mitted to the supporting walls, and plastered ceilings may be rendered defective on account of the vibration produced. It is therefore necessary to stiffen floors by providing cross *bracing* or *strutting* in continuous rows at intervals not exceeding 2 m apart. This is especially necessary when the joists are deep, as such have a tendency to twist or tilt sideways. As a rule, strutting is not necessary for ground floors which are partly supported by sleeper walls. There are two forms of strutting, *i.e.*, (1) herring bone and (2) solid.

1. *Herring Bone Strutting.*—This is undoubtedly the best form, and comprises pairs of inclined pieces of timber which are tightly fitted between the joists. The size of the pieces varies from 50 mm by 32 mm to 50 mm

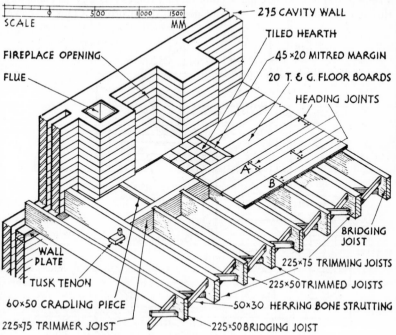

SCALE MM

275 CAVITY WALL
TILED HEARTH
45×20 MITRED MARGIN
20 T. & G. FLOOR BOARDS
HEADING JOINTS
FIREPLACE OPENING
FLUE
A
B
BRIDGING JOIST
WALL PLATE
225×75 TRIMMING JOISTS
TUSK TENON
225×50 TRIMMED JOISTS
60×50 CRADLING PIECE
50×30 HERRING BONE STRUTTING
225×75 TRIMMER JOIST
225×50 BRIDGING JOIST

T R I M M E D   F L O O R   A T   F I R E P L A C E
Fig. 58.

by 50 mm, and these are secured to the sides of the joists by a 70 mm nail at each end. This strutting is shown in Figs. 49, 50 and 58, and in the enlarged details in Fig. 59. Provided the walls are sufficiently strong, folding wedges are driven in at each end between the wall and the adjacent joist in line with the strutting (see A and B, Fig. 50 and Fig. 59); these are allowed to remain, as they increase the efficiency of the strutting. This form of strutting is still effective even if the joists shrink in the direction of their depth and thickness, for the depth shrinkage especially tends to reduce the inclination of the struts, with a corresponding tightening at the joints. Short saw cuts are sometimes made at the ends of the pieces (see J, Fig. 59) to prevent the nails splitting the timber; this reduces the holding power of the nails, and therefore cuts should not be made unless found to be necessary.

2. *Solid Strutting.*—The simplest form (and one which is often adopted in cheap work) merely consists of nailing short lengths of floor board in a continuous row between the joists at approximately 1·8 m intervals. *This is quite ineffective* on account of the shrinkage which occurs

in the thickness of the joists ; this shrinkage causes the struts to become loose, as their length is then less than the clear distance between the joists.

To make the strutting effective it is necessary to fix a long steel or wrought iron rod (of circular section varying from 12 to 25 mm diameter) through the whole of the joists and near to the strutting. This is shown

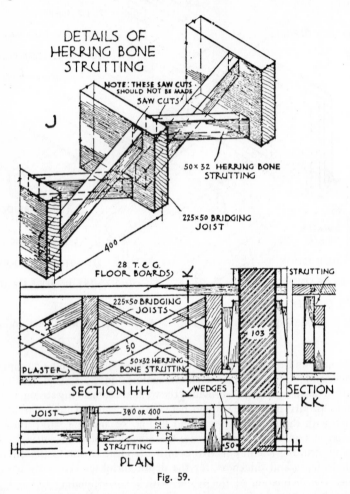

DETAILS OF
HERRING BONE
STRUTTING

NOTE: THESE SAW CUTS
SHOULD NOT BE MADE

SAW CUTS

J

50 × 32 HERRING BONE
STRUTTING

225×50 BRIDGING
JOIST

400

28 T. & G.
FLOOR BOARDS

STRUTTING

225×50 BRIDGING
JOISTS

103

50

50×32 HERRING
BONE STRUTTING

PLASTER

SECTION HH

WEDGES

SECTION
KK

JOIST

380 OR 400

52

32

STRUTTING

50

PLAN

Fig. 59.

at G (as an alternative to the herring bone strutting at H) and B, Fig. 50, and Fig. 60. The rod has a head at one end and its other end is threaded (to take a nut) ; it is passed through the holes bored (see " Brace and Bits," p. 213) through the middle of the joists. The nut is tightened by means of a spanner after the struts have been fixed, and again tightened before the floor boards are laid. The part vertical section through the two wall joists at T, Fig. 60, shows the ends of the rod. The object of the two washers is to prevent the head and nut from biting into the timber when the spanner is applied (see p. 199 and Fig. 153). This form of strutting (with rod) is now seldom adopted.

SUPPORTS FOR WALL PLATES.—It is occasionally necessary in buildings with 215 mm party walls[1] for the floor joists to be supported on the party wall. The Building Regulations do not permit such a wall to be penetrated by the joists because (a) of the danger of spread of fire from one building to the other and (b) the sound insulation of the wall would be impaired Hence, when the party walls are 215 mm thick it is necessary to provide either offsets

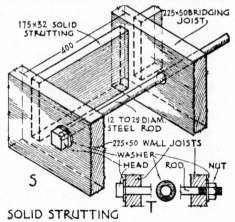

SOLID STRUTTING

Fig. 60.

(p. 76) or corbels. Thus, for a ground floor, 102·5 mm offset would be formed and the construction would resemble that shown in connection with the inner portion of the cavity wall at B, Fig. 36, except that the wall below the offset would be 327·5 mm thick and that above it would be 215 mm thick; alternatively, a sleeper wall bearing as shown in Fig. 35 could be provided. A 215 mm party wall must be corbelled for an upper floor, *i.e.*, the ledges on which the wall plates are supported are formed by projecting or oversailing courses. This construction is shown in Fig. 61, the wall plate, joists and boarding being indicated by broken lines. The section at L shows a 100 mm (or 115 mm) by 75 mm wall plate sup-

[1] A party wall is a division wall between two buildings belonging to different owners or occupied by different persons.

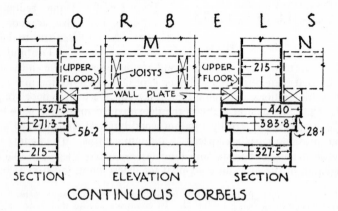

CONTINUOUS CORBELS

Fig. 61.

ported on a ledge formed by two projecting courses; a part elevation of this continuous corbel is shown at M. If a 215 mm party wall supports an upper floor at each side a corbel similar to L will be provided at both sides. The section at N, Fig. 61 shows a portion of a 327·5 mm party wall decreased to 215 mm for an upper storey; the reduction in thickness of walling, together with the corbel on each side, provides the required bearings for the wall plates.

An alternative form of construction sometimes adopted is shown in Fig. 62. This shows the wall plate supported on mild steel or wrought iron bars, called *corbel brackets*, which are from 75 to 100 mm wide by 10 mm thick by about 450 mm long with ends turned up 50 mm in opposite directions. After being painted, they are built by the bricklayer into the wall at from 750 mm to 900 mm apart.

DOUBLE FLOORS

As stated on p. 75 it is usual to limit the clear span of softwood bridging joists to 4·8 m and, therefore, when the width of a room exceeds this

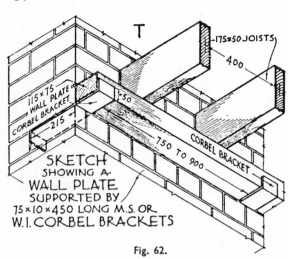

figure one or more relatively large members, called *binders*, are introduced to act as intermediate supports for the bridging joists. Such is called a *double floor*. The introduction of binders results in an economy in material, and the bridging joists, being reduced in size, are more convenient to handle. The binders are

Fig. 62.

spaced at from 1·8 to 3 m centres, and are placed across the shortest span in order that their dimensions may be kept down to a minimum.

Plan and sections showing a typical lay-out of a double floor are given in Fig. 63. The plan at A shows the floor divided into three bays by the provision of two 380 mm by 175 mm wood binders[1], and the 150 mm by 50 mm bridging joists are spaced at 380 mm centres. The factors which influence the sizes of joists (and binders) are stated on p. 74; whilst an explanation of the procedure adopted for determining the sizes is beyond

[1] For the reasons given on p. 105 wood binders are very seldom used in this country, steel beams being preferred. In countries having their own supply of cheap timber this construction is common.

the scope of this book, it should be stated that in obtaining these it was assumed that the superimposed load was 2·4 kN/m². The binders are shown supported on stone pads (see A, B and C, Fig. 63 and the enlarged detail in Fig. 65); these pads provide sound bearings and are effective in

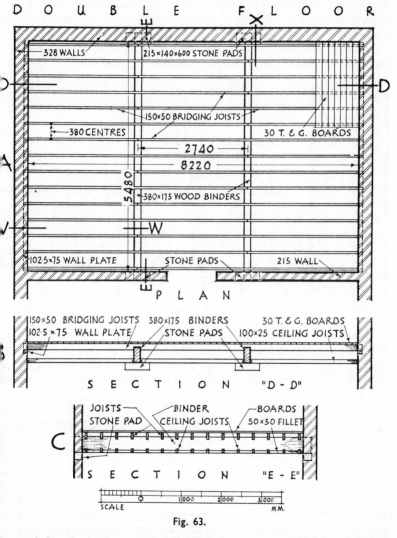

Fig. 63.

transmitting the loads to the brickwork below; 75 mm thick stone lintels are also sometimes built-in above the ends of the binders. The necessary circulation of air round the ends of the binders is assured if pockets are provided, as shown in Figs. 65 and 66 (H).

With a view to reducing the overall depth of the floor and effecting an economy in the brickwork of the adjacent walls, the bridging joists are cogged to the binders—see Fig. 39, Fig. 63 (B and C), Fig. 64 (B), Fig. 66 (H) and p. 78. The depth of the cogging (or cog) should not exceed two-thirds the depth of the bridging joists and their bearing need not exceed 25 mm. Whilst such sinkings do not much reduce the strength of the binders, provided the workmanship is sound and the joists are a tight fit, the cutting and notching of bearing timbers, such as binders, should be restricted as much as possible.

CEILING.—If the ceiling of the room is required to be flush with the soffit (underside) of the binders, the necessary construction is as shown

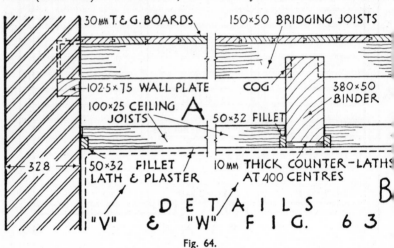

Fig. 64.

at B and C, Fig. 63 and Figs. 64, 65 and 66; 50 mm by 32 mm (or 25 mm) fillets are securely nailed to the sides of the binders (see B, Fig. 63, Fig. 64, Fig. 65 and H, Fig. 66) and plugged to the walls (see A, Fig. 64 and G, Fig. 66). As shown, the ends of the 100 by 50 mm ceiling joists are notched

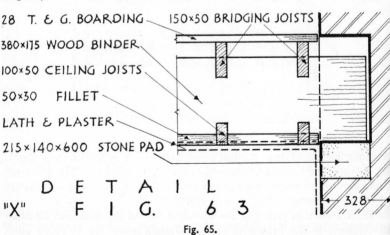

Fig. 65.

to these fillets and nailed. Plasterers' laths (or metallic lathing) are nailed to the ceiling joists (see G, Fig. 66). Short pieces of thick laths, called *counter-laths*, are nailed at 380 mm centres to the soffit of the binders (see B, Fig. 64, H, Fig. 66 and S, Fig. 70). Such provision should be made when timbers exceed 75 mm in thickness and so afford a proper key for the plaster; see A, Fig. 71, which shows the plaster that has been pressed through the spaces between the laths.

Several alternative details, showing the ceiling attached direct to the bridging joists, appear in Fig. 71 (see pp. 109-111).

STEEL BINDERS.—The size of the wood binders shown (380 mm by 175 mm) is much in excess of the normal stock sizes. Difficulty is often experienced in obtaining sound timber of such large size. This is one of the reasons why mild steel has largely superseded timber as a material for floor members such as binders. Mild steel, which is produced from iron ore is rolled into sections of several shapes. The section of steel members used in lieu of a wood binder is of I form; one type is the *rolled steel joist* (R.S.J.), another is the universal beam (U.B.). The top and bottom portions are called *flanges* and that connecting the flanges is known as the *web*.

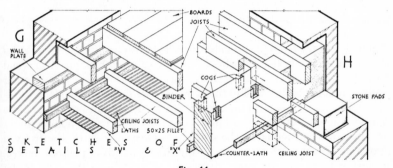

Fig. 66.

Steel beams are rolled to many sizes and are specified according to their overall depth, the width of the flanges and the weight per lineal metre. Thus, the steel binder shown in Fig. 67 is 203 mm by 102 mm by 25·3 kg; a larger steel beam or girder is shown in Fig. 69.

A detail incorporating a rolled mild steel beam as a binder instead of one in wood is shown in Fig. 67. As mild steel is considerably stronger than timber, the size of the steel binder is less than that of a wood binder. Thus, the 203 mm by 102 mm by 25·3 kg steel beam shown in Fig. 67 will support the same load as the 380 mm by 175 mm timber binder (compare Fig. 67 with B, 64). One result of this reduction in size is the corresponding decrease in the amount of walling, equivalent in this case to approximately two courses of brickwork (140 mm).

The section in Fig. 67 shows the steel binder with the bridging joists notched at its top flange and supported on 50 mm by 50 mm bearers which are secured to the web of the binder by bolts (see broken line) at about

600 mm centres; alternatively and preferably, deeper bearers may be employed in order that they may be supported on the bottom flange of the binder (similar to F, Fig. 71—see p. 111). In this detail (Fig. 67), unlike that at B, Fig. 64, the bridging joists are covered with plasterboard and

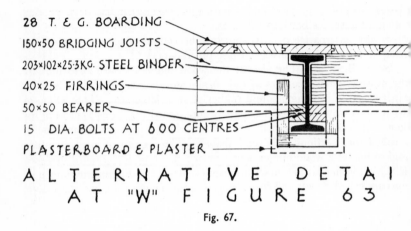

28 T. & G. BOARDING
150×50 BRIDGING JOISTS
203×102×25·3 KG. STEEL BINDER
40×25 FIRRINGS
50×50 BEARER
15 DIA. BOLTS AT 600 CENTRES
PLASTERBOARD & PLASTER

ALTERNATIVE DETAIL
AT "W" FIGURE 63

Fig. 67.

plastered, and the binder is provided with *firring* (or *cradling*). This firring consists of two vertical 40 mm by 25 mm (or 30 mm) thick pieces of wood nailed to the sides of each pair of timber joists and the bearers, and a similar bottom cross firring is nailed to them. There are several methods of forming the firring. Thus : (1) that shown in Fig. 67 indicates the cross piece nailed to the sides of the vertical pieces ; (2) at A, Fig. 69 the bottom pieces are sawn to length equal to the distance between the vertical strips and secured by driving nails through the latter into the ends of the soffit pieces, and (3) at B, Fig. 69 the cross strips are nailed through the ends of the vertical members (where there is a tendency for the joints to be forced open if the vertical strips shrink in their length and no space is left between the bottom flange of the steel beam and the cross strips). The plasterboard (fixed by the plasterer) is nailed to this firring, as shown at S, Fig. 70, and two coats of plaster are then applied.

Detail E, Fig. 71, shows a steel binder (see p. 111).

TRIPLE OR FRAMED FLOORS

This type of floor consists of three sets of joists, *i.e.*, bridging joists, binders and girders (see p. 71). Formerly, the binders and girders were of wood and the binders were framed or tenoned to the girders. Girders, like most binders (p. 105), are now made of mild steel. A triple floor may be adopted when the narrowest span exceeds 7·3 m and the live or superimposed load (p. 74) is relatively heavy.

The part plan of a triple floor capable of supporting a live load of

4·79 kN/m², is shown in Fig. 68. The various members are indicated in outline. Mild steel girders span the room at 2·54 m centres ; these are

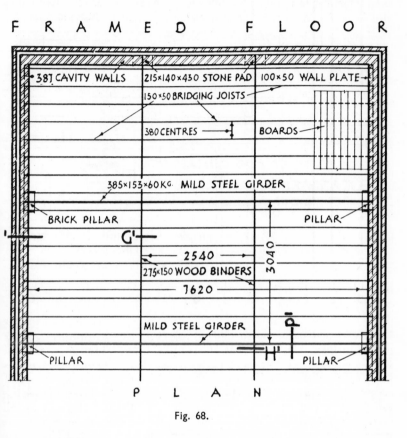

Fig. 68.

supported on hard stone pads bedded on brick piers (see also C, Fig. 69 and R, Fig. 70) formed on the 215 mm thick inner portion of each of the long 387 mm cavity walls to afford a good bearing. The girders support two binders at one-third points (2·54 m centres), and the binders carry the bridging and ceiling joists.

Details of this floor are illustrated in Fig. 69. That at F shows a cross section through the steel girder. The binders are notched over the top flange of the girder and are further supported on 90 mm (or 80 mm) by 90 mm by 8 mm (thickness) mild steel angles (of ⌐-section) secured to the web of the girder by welding or 20 mm dia. bolts at 380 mm centres (see also E, Fig. 69 and R, Fig. 70). These angles also support the 75 mm by 50 mm bearers (having their ends notched and nailed to the 50 mm by 25 mm fillets nailed to the binders—see E) which provide fixings for the ends of the vertical firrings. The methods of joining the latter at A and B

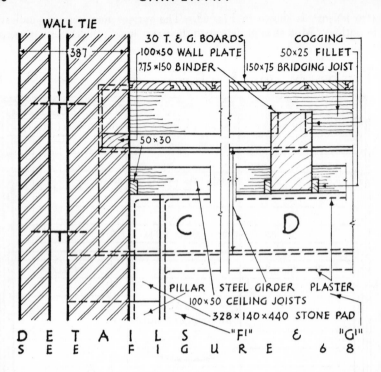

WALL TIE
387
30 T. & G. BOARDS
100×50 WALL PLATE
275 ×150 BINDER
COGGING
50×25 FILLET
150×75 BRIDGING JOIST
50×30

C   D

PILLAR   STEEL GIRDER   PLASTER
100×50 CEILING JOISTS
328×140×440 STONE PAD

D E T A I L S   "F¹" &   "G¹"
S E E   F I G U R E   6 8

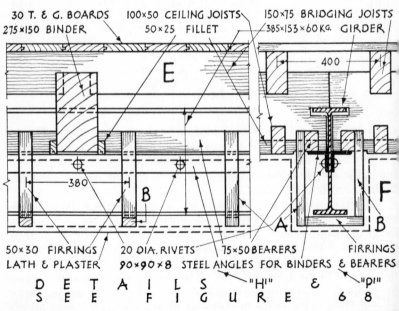

30 T. & G. BOARDS   100×50 CEILING JOISTS   150×75 BRIDGING JOISTS
275×150 BINDER   50×25 FILLET   385×153×60 KG. GIRDER

400

E

380

B   A   F
                 B

50×30 FIRRINGS   20 DIA. RIVETS   75×50 BEARERS   FIRRINGS
LATH & PLASTER   90×90×8 STEEL ANGLES FOR BINDERS & BEARERS   & BEARERS

D E T A I L S   "H¹" &   "D¹"
S E E   F I G U R E   6 8

Fig. 69

have been described on p. 106. The detail at E shows the part elevation of the girder, etc. and cross-section through a binder. The above construction is shown in the sketch s, Fig. 70. The details of the binders, bridging joists and ceiling joists are shown at C and D, Fig. 69, and are similar to those of the double floor. The detail at C also shows the bearing for the girder (see also sketch R, Fig. 70).

Each of these 275 mm by 150 mm wood binders may consist of two 275 mm by 75 mm joists, bolted together as shown at B and C, Fig. 71 (see pp. 110 and 111), the double row of bolts being staggered. The use of such stock sizes may be preferred if the larger single members are not available.

If mild steel is used instead of wood for the binders, calculations show that the size of the steel binders need only be 152 mm by 89 mm by 17·1 kg. Such steel members greatly simplify the details, as the 150 mm by 75 mm bridging joists would just be notched at both flanges of each steel binder, and a flush ceiling would result, as the plasterers' laths would be nailed direct to the joists and the ceiling joists omitted. (Note that timber laths are now seldom adopted, metallic lathing and plaster being preferred.) If the bridging joists are cut carefully and fitted tightly between the webs of the steel binders, no other fixing need be provided for the former.

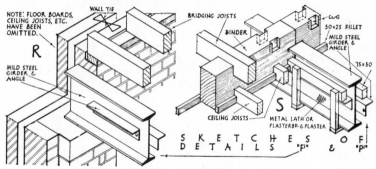

Fig. 70.

ALTERNATIVE DETAILS.—The double and framed floors detailed in Figs. 64, 65, 66, 69 and 70 have the ceilings flush with the soffits of the binders. Economy results if the ceiling joists are omitted and the plasterers metal lathing or plasterboard is applied to the bridging joists.

Detail A, Fig. 71, shows the bridging joists cogged to a *wrought* binder and the laths supporting the plaster nailed to the joists. Wrought means " work," and the work expended on the binder is the planing (see p. 209) on its sides and soffit necessary to bring them to a smooth or " dressed " condition. On the right the plaster is shown flush with the side of the binder ; as the binder may shrink and cause a gap between it and the plaster, a small rounded (or moulded) member, called a *bead*, is nailed to the binder to cover the edge of the plaster. Alternatively, as shown on the left, the binder may be grooved to receive the plaster. Note that this is a traditional finish no longer adopted, plasterboard or metal lathing have replaced timber laths.

Detail B, Fig. 71, shows the binder consisting of two undressed (not planed) deals (p. 16) which are bolted together at 600 mm intervals, the upper bolts (see Fig. 153) staggering with the lower. Two methods of covering the binder are indicated. That on the left shows a *lining* or *casing* of 16 mm dressed boarding secured to fillets nailed to the sides and soffit of the binder ; the thickness of the fillets varies according to the overall size and proportion of the binder desired. An alternative method is to fix the lining direct to the binder as shown on the right. The joints between the side and soffit lining boards may be either butt or square, *mitred* (adjacent edges being splayed at an angle of 45°), *tongued mitred* (the adjacent mitred edges being grooved to receive a narrow wood slip or tongue which is glued before being inserted in the grooves as shown in

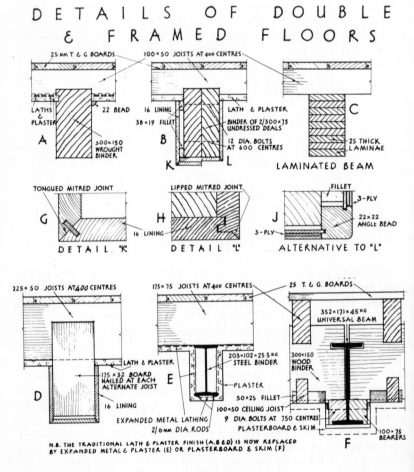

Fig. 71.

the detail at G) or, preferably, *lipped mitred* as shown at H. The tongued mitred and lipped mitred joints are adopted in good class work, as such are not likely to open.

The alternative detail at C, Fig. 71, is a laminated beam consisting of 25 mm thick timber laminæ glued together (see also p. 153).

A mock beam or binder is shown at D, Fig. 71. Short lengths of floor boards, or equivalent, are nailed to alternate joists and the lining, jointed as described above, is fixed to these. Whilst this construction does not strengthen the floor, it is sometimes adopted to divide the ceiling into bays.

The detail at E, Fig. 71, shows expanded metal lathing wrapped round the steel beam ; small rods below the bottom flange of the beam allow the plaster to penetrate the lathing here.

NOTE.—The covering or lining of binders and similar members is included in joinery and is referred to here, as it is associated with floor construction. To ensure tight-fitting joints between the members comprising the linings it is usual to prepare the latter on a joiners' bench (Fig. 161) ; the joint edges are fitted together, trued or planed where necessary to ensure a good fit along their entire length, glued and then cramped (secured by cramps, see p. 215, until the glue has set). The linings are then fixed to the floor members.

Detail F, Fig. 71, is alternative to that shown at F, Fig. 69 (see p. 107). The binders near their top edges are notched over the upper flange of the steel girder and are supported on continuous wood bearers, which latter are bolted to the web of the girder.

DOUBLE BOARDED FLOORS.—Double flooring is sometimes adopted for domestic and other buildings which require good class floors. As is implied, the floor covering is of two thicknesses. The floor usually consists of a first covering (or *sub-floor* or *counter-floor*) of 19 or 25 mm roughly sawn square edged boards *laid diagonally* across the wood joists to avoid their longitudinal joints coinciding with those of the boards above. The upper boards, usually of hardwood, are fixed to the lower boards at right angles to the joists. The sub-floor may consist of sheets of plywood (15 mm thick 5-ply up to 1·2 m wide and 2·44 m long, depending upon the spacing of the joists—see Chapter Two) instead of boards. Usually, any plastering to walls and ceilings is completed before the top flooring is fixed. This is one of the advantages of a double floor as, during the operation of plastering, portions of plaster falling on to an unprotected floor stain the timber and produce scratches on the surface when rubbed in with the feet (see p. 87).

Wood blocks (when not laid on a concrete sub-floor), plywood flooring and parquetry are fixed to wood counter-floors. These, together with boarding fixed to bearers (embedded in concrete or secured by floor clips) and plank floors, are described in the companion volume " Joinery."

CHAPTER FIVE

ROOFS

Terms. Single roofs, including flat, lean-to, double lean-to, couple, close couple
and collar roofs. Double roofs ; dormer window ; hips and valleys. Trussed
rafter roofs and built-up roof trusses, glued laminated portal frame ; hyperbolic paraboloid roof ; king post roof truss. Roof trimming. Flush, open,
closed and sprocketed eaves. Determination of bevels. Erection. Gutters.

THE following are some of the terms applied to timber roofs. They are
illustrated in Fig. 72 and subsequent drawings.

COVERING.—This is the external material fixed on a roof to protect
the building. The materials used for this purpose include : Slates (of
various sizes, but commonly 400 to 600 mm by 200 to 300 mm, and obtained
by splitting blocks of the laminated stone into slabs which are from 4 to 5
mm thick—see Figs. 99, 100, 102 and 106), plain tiles (burnt slabs of clay,
usually 267 mm by 165 mm by 13 mm and slightly curved longitudinally
—see Fig. 100), pantiles (slabs of clay, 360 mm by 250 mm by 16 mm,
moulded to a flat S in cross section, and burnt—see Fig. 113), lead (mostly
obtained in 215 cm wide rolls weighing 4 to 7 lb. per sq. ft.[1]—see Figs. 74, 75,
76, 120, 121 and 125), asbestos-cement corrugated sheets (90 to 300 cm by
6·5 to 9 cm, a very light covering especially suitable for the factory type of
building—see Fig. 113), zinc (thin sheets, usually 240 by 90 cm and laid
somewhat like lead to form a cheaper and inferior covering), copper (thin
sheets, 370 cm by 120 cm maximum size, forming a very durable covering),
aluminium (thin sheets of various sizes), stone slating (thicker and heavier
than ordinary slates of varying sizes), shingles (thin slabs of wood, usually
cedar, about 400 mm long and up to 350 mm wide, tapering from 10 to 3
mm thick), patent glazing (sheets of glass supported by lead covered wood,
steel or reinforced concrete bars), asphalt felt (in 90 cm wide rolls—see
Fig. 73), galvanized wrought iron corrugated sheets (140 to 300 cm by
60 cm) and thatch (bundles of reed or straw laid to a thickness varying from
30 to 40 cm).

SPARS OR COMMON RAFTERS.—Similar to joists but inclined. The
distance apart depends upon the covering material and is usually 380 or
400 mm centres for slates and tiles ; the *head* of a spar is the upper end,
and the *foot* is the lower end.

SPAN.—Usually taken to be the clear horizontal distance between
the faces of the walls supporting the roof. The *effective span* is the horizontal distance between the centres of the supports. The span of spars
is the inclined distance from support to support ; thus, in Fig. 90 the

[1] Lead is still made to this Imperial measure but specified as No. 4, 5, 6 and 7
grade lead.

span is the distance from ridge to purlin, purlin to purlin, and purlin to wall plate.

RISE.—The vertical height measured from the lowest to the highest points (see B, Fig. 72).

PITCH.—The slope or inclination to the horizontal expressed as rise ÷ span (see B, Fig. 72) or in degrees. It varies with the covering material in accordance with Table II.

TABLE II. ROOF PITCHES

Covering Material	Rise (mm) (in 100 mm run) (see B, Fig. 72)	Minimum Pitch	Angle (degrees)
Copper	1·25	$\frac{1}{160}$	$\frac{3}{4}$
Lead and zinc (excluding drips every 3 m run)	1·25	$\frac{1}{160}$	$\frac{3}{4}$
Asphalt felt, corrugated asbestos and iron sheets	10	$\frac{1}{20}$	$5\frac{3}{4}$
Slates, large . . .	40	$\frac{1}{5}$	$21\frac{3}{4}$
Slates, ordinary . . .	50	$\frac{1}{4}$	$26\frac{1}{2}$
Slates, small . . .	66·6	$\frac{1}{3}$	$33\frac{2}{3}$
Pantiles . . .	45	$\frac{1}{4.4}$	24
Shingles and patent glazing .	50	$\frac{1}{4}$	$26\frac{1}{2}$
Stone slating . . .	66·6	$\frac{1}{3}$	$33\frac{2}{3}$
Plain tiles and thatch . .	100	$\frac{1}{2}$	45
Interlocking tiles . . .	50·8	$\frac{1}{3.5}$	30

Roofs are often constructed with pitches much in excess of the above minima. For example, whilst lead is commonly used to cover flat roofs which have a minimum rise of 1·3 mm for 100 mm run, it is sometimes used as a covering to steeply pitched roofs. Copper is also adopted for high pitched roofs. Much depends upon the appearance required. Roofs pitched at 45° are not of satisfactory appearance, but those having an angle between 50° and 60° look well ; compare the roof shown in Fig. 84 (a 45° slope) with that in Fig. 90 (a 55° pitch).

WALL PLATES.—These receive the feet of spars. They are usually 100 mm by 75 mm and are bedded and jointed as described on pp. 75-79.

EAVES means " edge ", and the eaves of a roof is its lower edge. The eaves may finish flush with the outer face of the wall, when it is known as a *flush eaves* (see Figs. 73 and 99), or it may project as shown in Figs. 90, 93, 100, 101, 102, etc. When the feet of the spars are exposed as in Figs. 100 and 101 they form an *open eaves* ; when the feet are covered as shown in Figs. 102 and 103, a *closed eaves* results. A *fascia* (or *fascia board*) is the thin piece of wood fixed to the feet of the spars (see Figs. 102 and 105). The under portion of an overhanging eaves is called the *soffit* ; *soffit boards* are shown in Figs. 102 and 103, and the crosspieces of wood included in the latter figure to which these boards are nailed are called *soffit bearers*. The lower portion of a roof is sometimes tilted so as to improve its appearance; this is done by nailing short pieces of timber (*sprockets*) to the feet of the spares to give a *sprocketed eaves* (see Figs. 102 and 104).

RIDGE OR RIDGE PIECE.—This is fixed at the highest point to receive the heads of the spars.

HIP is the line produced when two roof surfaces intersect to form an external angle which exceeds 180°. A *hipped end* is a portion of roof between two hips (see A, Fig. 72). The timber at the intersection is called a *hip rafter*, and the foot of this rafter is usually fixed to a horizontal cross member called a *dragon beam* which is secured at one end to an *angle tie* (see Figs. 90 and 91). A hip rafter supports the upper ends of short spars (called *jack rafters*—see below) and it may be required to carry the ends of *purlins* (see below).

VALLEY is formed by the intersection of two roof surfaces having an external angle which is less than 180° (see A, Fig. 72) and the timber member at the intersection is called a *valley rafter* (Fig. 119). Jack rafters and sometimes purlins are nailed to a valley rafter.

JACK RAFTERS.—These are short spars which run from a hip to the eaves or from a ridge to a valley (A, Fig. 72 and Fig. 90).

PURLINS are horizontal timbers which provide intermediate supports to spars, and are supported by walls, hip and valley rafters, and roof trusses (see Figs. 86, 90 and 93).

ROOF TRUSSES are structures formed of members framed together. They support purlins in the absence of crosswalls. The *built-up* roof

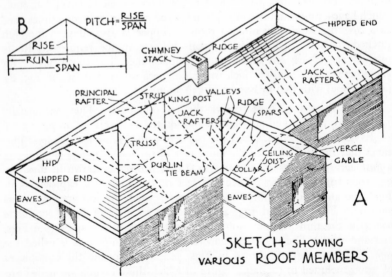

Fig. 72.

truss is one form of this type of structure ; it is detailed in Fig. 113. The queen post roof truss, formerly used for large spans, is now obsolete as is the king post truss (Fig. 116).

BOARDING OR SARKING consists of 25 mm (nominal thickness) boards which are nailed to the *backs* (upper edges) of spars, and to which slates and other roofing materials are secured (p. 112 and Figs. 98, 99 and 102).

BATTENS.—As stated on p. 16, these are wood members of small scantling and are used as fixings for slates, tiles, etc. (see Figs. 100 and 101). They are usually fixed by the slater or tiler.

CLASSIFICATION.—This is similar to the classification of boarded and joisted floors stated on p. 71, *i.e.* :—

(a) *Single Roofs.*—This class consists only of spars which are nailed to the ridge and wall plates.

(b) *Double or Purlin Roofs.*—In this type additional members, called purlins, are introduced to support the spars.

(c) *Trussed Rafter Roofs.*—This comprises light trusses formed by framing together spars and ceiling joists with intermediate members. They have almost entirely replaced purlin roofs for domestic work.

(d) *Triple or Framed Roofs.*—This class has three sets of members, *i.e.*, spars that are partially supported by purlins, which latter are carried by trusses.

These roofs have boarding (nailed to the backs of the spars) and/or battens to which the roof covering is fixed.

SINGLE ROOFS

The various forms of this class are : (1) flat, (2) lean to, (3) double lean-to, (4) couple, (5) close couple and (6) collar roofs.

Before considering each in detail it should be explained that the sizes of the spars specified on the various drawings must not be taken to be the most economical in all cases, as the sizes depend upon the weight of the covering material, the distance centre and centre of the spars, and the timber employed, in addition to the span of the spars. The weight of the covering material varies considerably :—

TABLE III. WEIGHTS OF ROOF COVERING

Material	Weight (kg/m²)	Material	Weight (kg/m²)
Zinc and copper . . .	2·4	Lead (including rolls) .	33·5
Asphalt felt . . .	3·6	Thatch . . .	33·5
Shingles, cedar . . .	7·2	Interlocking tiles . .	36·0
Corrugated iron . . .	12·0	Slates	43·0
Boarding, 1-in. thick .	14·4	Pantiles . . .	48·0
Corrugated asbestos-cement .	16·8	Plain tiles . . .	62·2
Patent glazing (steel) . .	29·0	Stone slating . .	86·2

Tables are provided in the Building Regulations which give the size of spar according to its span, pitch and load. The spars most commonly used are 50 mm thick and for average good class work are of redwood or Douglas fir.

1. FLAT ROOF.—(See Fig. 73). The construction is similar to that

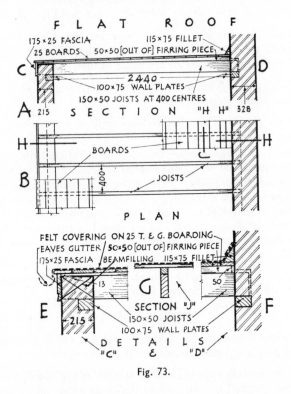

Fig. 73.

of a floor, the joists being spaced at 380 or 400 mm centres and covered with 25 mm thick boarding. The covering material may be lead, copper, zinc, soft aluminium sheet or asphalt felt.[1] The upper surface must be inclined sufficiently to throw off the water (see Table II, p. 113). If the under surface is not required to be level, the inclination is obtained by

---

[1] Roofs of temporary buildings are usually covered with one layer of felt. In this example, however, it is assumed that the covering consists of *three* layers of asphalt felt in 90 cm wide rolls (see p. 80) and the *minimum* pitch need therefore be only that for lead (see Table II, p. 113). The first layer, laid direct on the boarding, is lapped 75 mm at the joints, with adhesive between, and nailed at 75 mm intervals along the joints. Hot bitumen adhesive is applied over this layer, a second layer of felt is laid with 75 mm lapped joints (not nailed), bitumen adhesive is brushed over it, and a third layer of felt is laid as described. A top coat of hot bitumen adhesive is applied, and grit (slate granules) is now rolled into the adhesive to protect the felt from the action of the sun. As indicated by broken lines at F, Fig. 73, the layers of felt are continued over a triangular fillet fixed in the angle, and their edges are covered with a lead flashing.

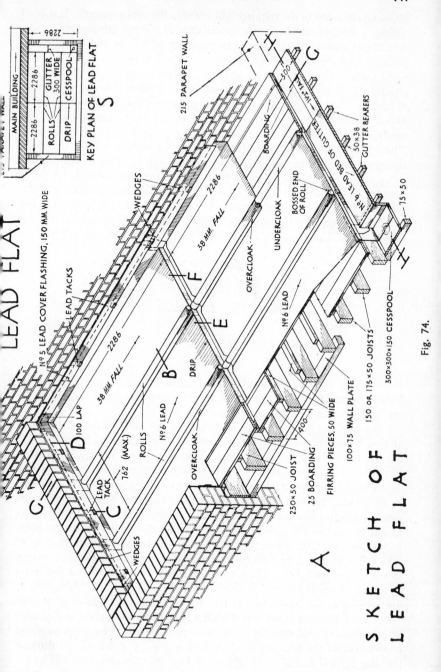

KEY PLAN OF LEAD FLAT

MAIN BUILDING

GUTTER 300 WIDE

ROLLS

DRIP

CESSPOOL

2286

2286

2286

215 PARAPET WALL

LEAD FLAT

Nº 5 LEAD COVER FLASHING, 150 MM WIDE

LEAD TACKS

WEDGES

2286

38 MM FALL

2286

38 MM FALL

100 LAP

762 (MAX)

ROLLS

Nº 6 LEAD

LEAD TACK

WEDGES

OVERCLOAK

DRIP

Nº 6 LEAD

BOARDING

UNDERCLOAK

OVERCLOAK

BOSSED END OF ROLL

Nº 6 LEAD BED OF GUTTER

50×38 GUTTER BEARERS

75×50

230×50 JOIST

25 BOARDING

FIRRING PIECES, 50 WIDE

100×75 WALL PLATE

150 OR 175×50 JOISTS

300×300×150 CESSPOOL

SKETCH OF LEAD FLAT

Fig. 74.

inclining the joists to the required fall towards the eaves gutter. If a level ceiling is wanted, the fall may be obtained by either tapering the joists with the top edge of each sloped to the required fall, or alternatively, the joists of uniform depth may be fixed level and a small tapered piece of wood nailed on top of each. The latter method is usually adopted, as it is effective and economical ; it is shown in the section at A and the details at E, G and F, Fig. 73. These tapered pieces are called *firrings* or *firring pieces*. As shown at G, they are the same width as the joists, and in this case the depth varies from a maximum of 50 mm (see F) to 13 mm at E. The boarding is preferably tongued and grooved, and is nailed on top of the firrings ; the boarding should be dressed smooth in order to remove any sharp edges which may cause damage to the covering material. A fascia board is nailed to the ends of the joists to provide a suitable finish (p. 138). The detail at E shows a flush eaves (see pp. 113 and 138) ; alternatively, the eaves may project by continuing the ends of the joists beyond the external face of the wall and fixing a fascia and soffit board to them (as in Fig. 102).

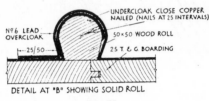

DETAIL AT "B" SHOWING SOLID ROLL

Fig. 75.

If the ceiling is to be plastered, it is advisable to provide a row of herring bone strutting (p. 99) along the centre between the joists, to stiffen the construction and prevent plaster cracks.

The construction of a large lead-covered flat roof is beyond the scope of this book, but in order to give some idea of a typical lay-out a lead flat has been illustrated in Fig. 74. The size of each sheet of lead must be restricted to prevent excessive expansion and contraction of the material. Accordingly, the flat is shown divided into two by a *drip* and each half is subdivided by two *rolls* (see Fig. 75). The boarding is given a fall towards a gutter (see the key plan at S and the sketch at A). As the boarding should be laid in the direction of the grain of the timber to ensure satisfactory drainage, the joists supporting it are laid across the shortest span. The fall is obtained by fixing rectangular (not tapered in section) firrings to the tops of the joists. These firrings increase in depth from a minimum of 13 mm at the lower joist to a maximum at the upper end ; deep firrings are avoided at the upper half of the flat by using deeper joists as shown. A section through a wood roll (Detail B, Fig. 74) showing the nailed edge of one sheet of lead covered by the free edge (to allow for expansion and contraction) of the adjacent sheet is given in Fig. 75, and a section through the drip (Detail E, Fig. 74) is shown in Fig. 76. As shown at A, Fig. 74, the gutter is constructed of 50 mm by 38 mm cross pieces, called *gutter bearers*, at 400 mm centres ; these are fixed at different levels to give the necessary fall to the boarding which is nailed to them. These bearers are supported by the wall at one end and by a 40 mm thick longitudinal fillet

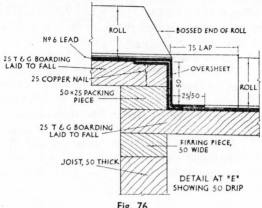

Fig. 76.

nailed to the side of the lower joist. The lower end of the gutter is finished with a woodlined box, called a *cesspool*, which is holed in the centre to receive the outlet pipe (see Fig. 124). Gutters are also described on pp. 162-165.

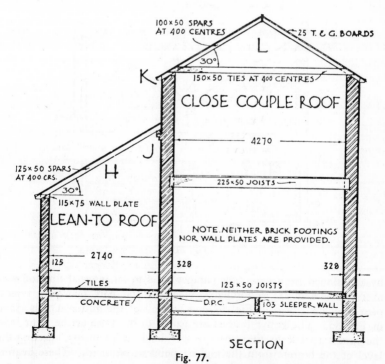

Fig. 77.

2. LEAN-TO ROOF.—This is the simplest form of pitched roof and consists of spars which are generally inclined at 30° against a wall when the covering is slates. A section through a building, part of which is covered by a lean-to roof, is shown at H, Fig. 77. An enlarged detail of J is shown in Fig. 78, where the wall plate is supported

by two brick corbel courses. Alternatively, metal corbel brackets as shown in Fig. 62 may be adopted. A cheaper method consists of nailing the upper ends of the spars to a continuous 75 mm by 50 mm *wall piece* or *pitch plate* which is plugged (p. 96) with its 75 mm face next to the wall. The construction of the eaves may be similar to the open eaves detailed in Fig. 100, except that there is no horizontal tie. The spars are V-shaped notched at both ends and fitted to the wall plates; such is one form of a *birdsmouthed joint*; another form is shown in Fig. 103. The depth of each notch shown in Figs. 78 and 99 should not exceed one-third that of the spar. Notching the spars counteracts the tendency for them to slide downwards. The eaves may be either flush (Fig. 99) or closed (Fig. 102).

3. DOUBLE LEAN-TO, PENT or V-ROOF.—Pent means penned or closed in, and this form consists of two lean-to roofs which are enclosed

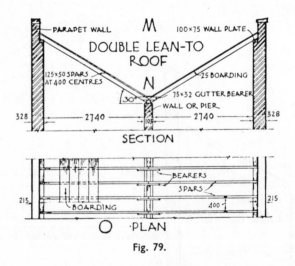

Fig. 79.

by and sloped from two outer parapet walls to an internal division wall over which a gutter is formed. A section and part plan of a typical pent roof are shown in Fig. 79, and a detail of the gutter at N is given in Fig. 80. The gutter bearers are nailed to the sides of the spars, and the necessary fall to the gutter is provided by gradually lowering the level of the bearers from the highest point as required. These bearers

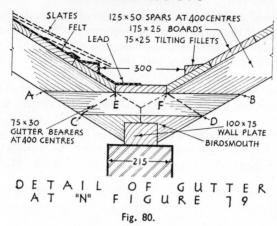

SLATES
FELT
LEAD
125 × 50 SPARS AT 400 CENTRES
175 × 25 BOARDS
75 × 25 TILTING FILLETS
300
A
E
F
B
75 × 30
GUTTER BEARERS
AT 400 CENTRES
C
D
100 × 75
WALL PLATE
BIRDSMOUTH
215

D E T A I L   O F   G U T T E R
A T   "N"   F I G U R E   7 9

Fig. 80.

C O U P L E   R O O F

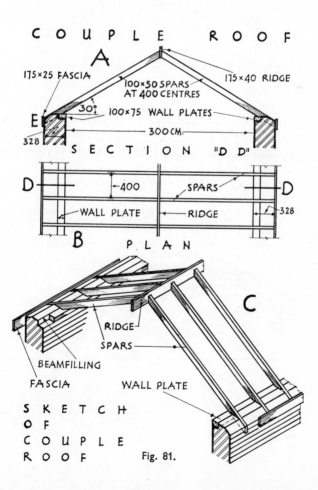

A
175 × 25 FASCIA
175 × 40 RIDGE
100 × 50 SPARS
AT 400 CENTRES
E
30°
100 × 75 WALL PLATES
300 CM.
328

S E C T I O N   "D D"

D
400
SPARS
D
WALL PLATE
RIDGE
328
B
P L A N

C
RIDGE
SPARS
BEAMFILLING
FASCIA
WALL PLATE

S K E T C H
O F
C O U P L E
R O O F
Fig. 81.

may be shaped as shown by full lines, ABCD (Fig. 80), but a saving in timber results if they are cut to the shape EFCD. The small *tilting fillet* at each side of the gutter is continuous; its purpose is to tilt the slates slightly. Instead of a division wall, the lower ends of the spars may be secured to a beam running parallel to the main walls.

The double lean-to roof is expensive because of the extra walling required, and the gutter is a potential weakness; hence it is seldom adopted.

4. COUPLE OR SPAN ROOF (see Fig. 81).—This is so called as each pair or couple of spars is pitched against each other and nailed at the upper ends to the ridge. A detail at the ridge is given in Fig. 82. The flush

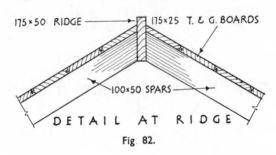

Fig. 82.

eaves is detailed in Fig. 99; alternatively, a projecting eaves of the type shown in Fig. 100 or Fig. 102 may be preferred. The shaping of the ends of the spars is described on p. 148. This roof is of bad design, as it has a tendency to spread at the feet and thrust out the walls. *It is not recommended*, but if used it should be restricted to a span not exceeding 3·6 m unless the walls are exceptionally thick.

5. CLOSE COUPLE ROOF.—This form of roof is illustrated in the section at L, Fig. 77, and the isometric sketch, Fig. 83. It is a great

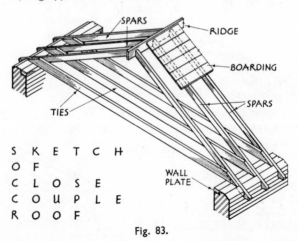

Fig. 83.

improvement on the couple roof, as each couple of spars is closed by a horizontal tie—hence the name—and it conforms with sound principles of construction. This tie is connected to the feet of the spars and prevents them spreading outwards. The best connection between a tie and a spar is the *dovetail halved joint* which is detailed in Fig. 85 and described on pp. 124-125. In cheaper work, however, the ties are simply spiked to the sides of the spars (and wall plates), as indicated in Fig. 83.

A plastered ceiling is often formed on the underside of the ties, which are then called *ceiling joists*. Such joists, when they exceed 3·6 m in length, should have vertical *hangers* nailed at the side to every third or fourth spar and to a horizontal *runner* which is laid on top of and nailed to the joists ; the hangers may be 75 mm by 38 mm or 50 mm by 44 mm (or short pieces of floor boards), and the runners are usually of 75 mm by 50 mm stuff. Such construction (see Fig. 94 and p. 132) prevents the sagging of the ceiling joists and cracking of the plaster.

The span of this type of roof should be restricted to 4·8 m unless the size of the spars and ties is increased (or the latter are supported by hangers), when the span may be increased to 6 m.

TABLE IV. SIZES OF TIES

Maximum Span (m)	Size (mm)	Maximum Span (m)	Size (mm)
2·4	100 by 50	4·25	175 by 50
3	125 by 50	5	200 by 50
3·65	150 by 50	5·5	225 by 50

If hangers or struts are used for spans of 3·65 m and upwards, the depth of the ties may be halved.

The detail of the open eaves at Fig. 77 is shown in Fig. 100, and an alternative closed eaves may be similar to that in Fig. 102.

6. COLLAR ROOF (see Fig. 84).—This somewhat resembles the close couple roof, except that the horizontal ties are now fixed at a higher level, and are called *collars*. The latter may be placed at any height between the wall plates and half-way up the roof, the broken lines at A indicating the position when at the maximum height. Clearly, the lower the collar the more effective it becomes in preventing the spars from spreading and causing damage to the walls. It follows, therefore, that the close couple roof is stronger than the collar type, but the latter permits an increase in the height of the room below. Thus, as shown by broken lines in the section, the plastered ceiling may be formed on the underside of the collars and the lower part of the spars. This economises in walling, for, if the same height of room was required and a close couple roof employed, the walls would have to be increased in height to the level of the collar.

The ends of the collar are commonly spiked to the sides of the spars and, therefore, appear like the ends of the ties shown in Fig. 83. In better class work, however, the *dovetail halved joint* (p. 125) is adopted. This is shown in Fig. 84 (compare the appearance of the ends in the axonometric sketch at B, Fig. 84, with those in the isometric sketch, Fig. 83. Details

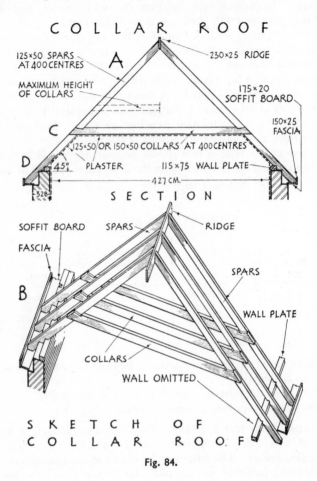

COLLAR ROOF

125×50 SPARS AT 400 CENTRES

230×25 RIDGE

MAXIMUM HEIGHT OF COLLARS

175×20 SOFFIT BOARD

150×25 FASCIA

125×50 OR 150×50 COLLARS AT 400 CENTRES

45°   PLASTER   115×75 WALL PLATE

427 CM.

32B

SECTION

SOFFIT BOARD   SPARS   RIDGE

FASCIA

SPARS

B

WALL PLATE

COLLARS

WALL OMITTED

SKETCH OF COLLAR ROOF

Fig. 84.

of the joint at C, Fig. 84, are given in Fig. 85. The latter shows that on one side of the spar a 13 mm sinking is formed with its upper edge dovetailed. The end of the collar is also checked out 13 mm on one side by a similar amount (13 mm), and the remainder of the thickness of the collar is dovetailed along the upper edge so that when the collar is fitted to the spar it will be housed to the extent of 13 mm (see section B'B'). The collar is then spiked to the spars. Alternatively, the depth of the sinking in the side of the spar is increased to 25 mm and the end of the collar is checked out by a similar amount; hence, when the collar is assembled both of its sides at each end are flush with those of the spar. The inclined abutments of the collar (which should fit tightly against the underside of the spar) tend to prevent the spars from sagging; the joint is also effective in resisting any tendency for the spars to spread because of the top *shoulder* (edge)

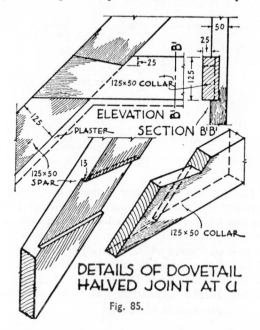

ELEVATION

SECTION B'B'

DETAILS OF DOVETAIL
HALVED JOINT AT C

Fig. 85.

of the collar bearing on the upper edge of the dovetailed notch formed on the spar.

The sizes of the collars should conform to those of the ties in Table IV (the "maximum span" being the clear length of the collar). It is not economical to adopt the collar type of single roof for spans exceeding 4·8 m.

Any of the single roofs may be given a different pitch to that indicated in each of the illustrated examples. It should also be pointed out that, whereas boarding is shown in some of the details, this is often omitted when battens (see Fig. 102 and p. 115) are used. The use of boarding helps to insulate the roof; but a better form of insulation is provided by putting an insulating quilt over the ceiling.

DOUBLE ROOFS

These are also known as *purlin roofs*, as purlins (p. 114) are introduced to provide intermediate supports to the spars. Purlins are necessary for roofs with spans of 5·5 m and upwards, otherwise the spars would need to be increased to an uneconomical size. The preferred maximum span (inclined) of spars is 2·45 m; this should be reduced to 1·8 m for small

pitched roofs with heavy covering. Sufficient purlins permit the use of
comparatively small (and therefore light) spars which are readily handled
on the job.

*All the single roofs already described may be adapted as double roofs by
the provision of one or more sets of purlins.*

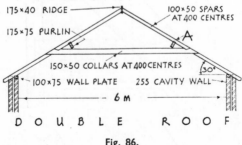

Fig. 86.

A simple example of a double roof is shown in Fig. 86. It is of the
collar type. A purlin is provided at each side to support the spars; the
latter are 100 mm by 50 mm, but if relatively thin slates are used as a
covering material 75 mm by 50 mm spars will suffice, as the clear span
(from wall plate to purlin and purlin to ridge) is only 1·75 m. The spars
are nailed to the wall plates, purlins and ridge, and to reduce any tendency
for the rafters to slide downwards they are cogged (see p. 78) 25 mm over
the purlins, in addition to birdsmouthing their lower ends to the wall plates.
Fig. 87 shows the spars cogged to a purlin.

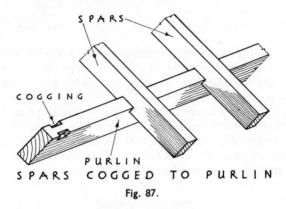

Fig. 87.

Purlins are supported by cross division walls of bedrooms, etc. (which
are carried up to the underside of the purlins. In the absence of cross
walls or partitions, it is usual to provide trusses (p. 149) to limit the
unsupported length of purlins to 4·8 m, as purlins exceeding this length are
not economical.

TABLE V. SIZE OF PURLINS

Span (m)	Maximum inclined distance apart (m)		
	1·8	2·3	2·75
1·8	115 mm by 75 mm	125 mm by 75 mm	150 mm by 75 mm
2·4	150 mm by 75 mm	175 mm by 75 mm	175 mm by 100 mm
3	175 mm by 75 mm	175 mm by 100 mm	200 mm by 100 mm
3·65	225 mm by 100 mm	225 mm by 115 mm	225 mm by 125 mm
4·25	225 mm by 125 mm	250 mm by 115 mm	250 mm by 100 mm
5	250 mm by 100 mm	250 mm by 125 mm	250 mm by 150 mm

Purlins may be placed normal (right angles) to the spars, as shown in Figs. 86 and 90, or vertically, as in Figs. 92 and 93. A level bearing on any cross walls is provided when the purlins are vertical.

Joints in long lengths of purlins are best arranged to coincide with and overlap at the wall supports as in Fig. 88. Sometimes, as indicated by the broken lines at N, stone pads are introduced at the supports to effectively distribute the weight on to the walls. Overlapping is not permitted if the division walls are party walls ; the latter, like joists (see p. 101), must be corbelled out to receive the ends of purlins (see p. 148).

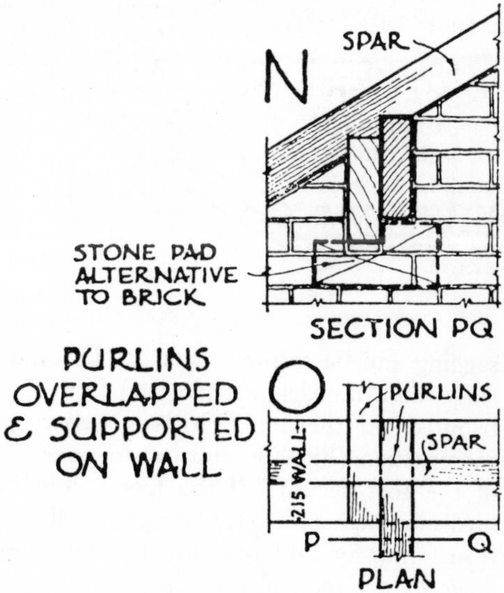

Jointing, known as *scarfing* or splicing, is resorted to when a purlin has to be lengthened. A *splayed* or *raking scarfed joint* is the

PURLINS OVERLAPPED & SUPPORTED ON WALL

Fig. 88.

best form for purlins. This is illustrated in Fig. 89 ; the length of the joint is from two to two and a half times the depth of the purlin ; right-angled cuts are made at the ends of the splayed portion : three or four 12 mm diameter bolts make the joint rigid ; a mild steel or wrought iron strap should be fixed at the underside of the joint (see sketch). This joint is also used for lengthening a ridge, where the length need only be one and a half times the depth ; a metal strap is not required, and long nails are used instead of bolts.

FISHING is an alternative form to scarfing. A fished joint is formed by butting the two squared ends of the timber together and connecting them by means of two metal (or wood) plates (one top and bottom) which are bolted as for a scarfed loint. The length of the plates equals four times the depth of the jointed member, and if wood plates are used their thickness should equal one-quarter this depth. This is a suitable joint for long struts.

Another example of a double roof of the collar type is shown in the section and part plan in Fig. 90. A hipped end (p. 114) has been introduced so as to illustrate the application of hip rafters and the construction involved. The spars are inclined at 55° (see p. 113) and two purlins are provided at each side to support the spars. At the hipped end the spars are cut short (when they are called *jack rafters*) and the heads are fitted (p. 143) and nailed to the hip rafters. The purlins have their ends shaped (p. 146) to make a tight fit with the sides of the hip rafters to which they are securely nailed. The ends of purlins are shaped and fixed in a similar manner to valley rafters. The collars, which are usually fixed to the spars immediately below the lower set of purlins, are shown dovetail halved jointed to the spars (see Fig. 85).

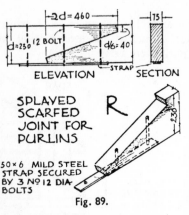

SPLAYED
SCARFED
JOINT FOR
PURLINS

50 × 6 MILD STEEL
STRAP SECURED
BY 3 N° 12 DIA.
BOLTS

Fig. 89.

Hip rafters are often required to carry relatively heavy loads from the purlins. They must therefore be of sufficient strength to prevent sagging and be securely fixed. The head of each rafter is nailed to the ridge, and in order that the load from the rafter shall be adequately distributed on the walls, it is necessary to employ a special form of construction (Figs. 90 and 91) to receive its foot and to make the angle of the roof secure; if the feet of hip rafters were, like spars, simply birdsmouthed and spiked to the wall plates, the concentrated inclined thrust may be sufficient to push out the corners of the building. An *angle tie* or *brace*, placed diagonally across the corner, is notched to the wall plates, and, to counteract the thrust, these notches should be dovetailed, as shown by the broken lines on the plan H, Fig. 91. The wall plates are half-lapped for the same reason and their ends project for about 75 mm. This angle tie carries one end of a beam, called a *dragon* (or *dragging*) *beam*, which is the chief support for the hip rafter; this beam is tusk tenoned to the angle tie and single cogged over the wall plates. The foot of the hip rafter is connected to the dragon beam by means of an *oblique tenon* joint and bolt as shown, or by a *bridle joint* as illustrated at E and J, Fig. 116. After fixing the hip rafter, the whole of the framing is made rigid by driving down the wedge of the tusk tenon. For lowly pitched roofs, and where the eaves is not sprocketed (pp. 139-140), the foot of the hip rafter is sometimes projected beyond the outer face of the wall to the line of the projecting feet of the spars. In this case the

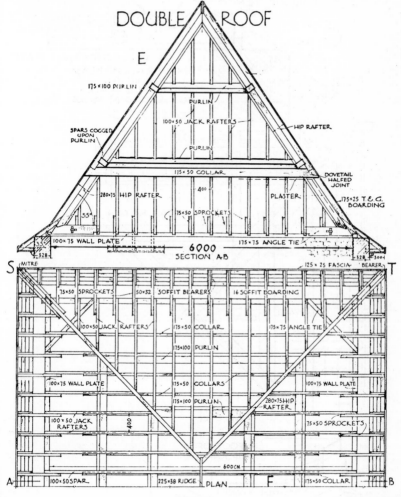

Fig. 90.

rafter is notched over and is tenoned nearer to the outer end of the dragon beam.

The eaves details are described on p. 140. The determination of the bevels of hips and purlins is described on pp. 144-147.

This type of roof, employing purlins and collars, is sound and economical construction. It is particularly suited for spans which do not exceed 7 m.

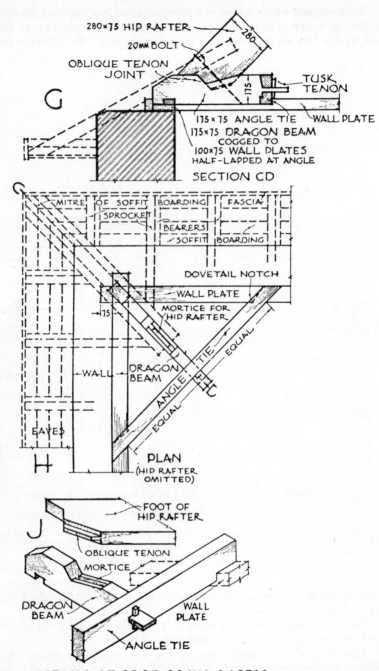

280×75 HIP RAFTER

20mm BOLT

OBLIQUE TENON
JOINT

G

TUSK
TENON

175×75 ANGLE TIE   WALL PLATE
175×75 DRAGON BEAM
COGGED TO
100×75 WALL PLATES
HALF-LAPPED AT ANGLE

SECTION CD

MITRE OF SOFFIT   BOARDING   FASCIA
SPROCKET
BEARERS
SOFFIT BOARDING

DOVETAIL NOTCH

WALL PLATE

MORTICE FOR
HIP RAFTER

75

WALL

DRAGON
BEAM

ANGLE TIE

EQUAL

EQUAL

C

EAVES

H

PLAN
(HIP RAFTER
OMITTED)

FOOT OF
HIP RAFTER

J

OBLIQUE TENON

MORTICE

DRAGON
BEAM

WALL
PLATE

ANGLE TIE

DETAILS AT FOOT OF HIP RAFTER

Fig. 91.

Another type of double roof is shown in Fig. 92. It is similar to the close couple roof (p. 122), with the addition of purlins. The spars are pitched at 30° (depending upon the covering material and design desired), birdsmouthed to the wall plates, notched over one pair of purlins (which are placed vertically as an alternative to those shown in Figs. 86 and 90) and spiked to the ridge. The ceiling joists or ties are secured to the wall plates and the feet of the spars as already described, and as these are partly supported by two sets of hangers and runners, the size of these joists need only be 100 mm by 50 mm or 125 mm by 50 mm (Table IV, p. 123), depending upon the weight of the covering material. The hangers and

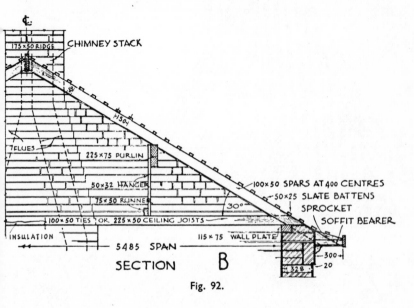

Fig. 92.

runners have been described on p. 123. Sometimes the runners are notched over the ceiling joists to give these additional rigidity. It is important that the lower ends of the hangers are not secured to the runners until *after* the slates or other covering material have been fixed otherwise the weight of this material may cause the spars to sag slightly, which in turn would depress the ceiling joists through the hangers.

A view of the general construction of this roof is shown in the isometric sketch, Fig. 93. The eaves is detailed in Fig. 106 and described on p. 142.

Fig. 94 shows another example of a double or purlin roof of good appearance. Because of its steep pitch (55°) it is often employed for attic bedrooms. It has spars, supported by two sets of purlins and lower or principal collars; the spars and these collars are spaced at 400 mm centres. In addition, there is a second set of collars supporting the upper pair of purlins; these collars are provided at every third, fourth or fifth pair of spars, to which they are securely nailed and preferably dovetail halved jointed (p. 124). The portion of roof between the collars is triangulated by braces or struts, birdsmouthed to the purlins, and a central runner

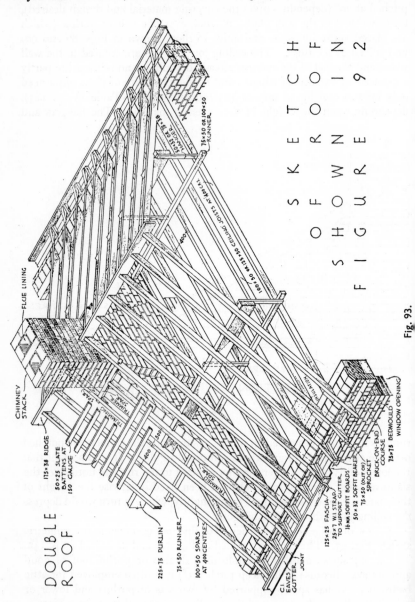

SKETCH OF ROOF SHOWN IN FIGURE 92

DOUBLE ROOF

FLUE LINING

CHIMNEY STACK

115×38 RIDGE

50×25 SLATE BATTENS AT 150 GAUGE

75×50 OR 100×50 RUNNER

100×50 HANGER

75×50 OR 125×50 CEILING JOISTS AT 400 C/S

400

TRIMMING SPAR

TRIMMED SPAR

400

400

225×75 PURLIN

75×50 RUNNER

100×50 SPARS AT 400 CENTRES

TRIMMER SPAR

TRIMMED SPAR

125×25 FASCIA

25×7 W.I. STRAP TO SUPPORT GUTTER

16 mm SOFFIT BOARDS

50×32 SOFFIT BEARER

75×50 (CUT OF) SPROCKET

BRICK-ON-END COURSE

75×75 BEDMOULD

WINDOW OPENING

TRIMMED SPAR

C.I. EAVES GUTTER.

JOINT

Fig. 93.

nailed to the ceiling joists and placed immediately over a stout stoothed partition (p. 166). Hangers and side runners are provided as intermediate supports to the ceiling, the former being spiked to the purlins and runners.

DOUBLE ROOF

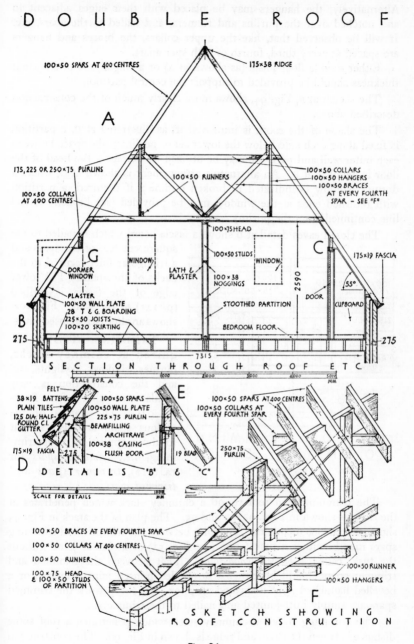

100×50 SPARS AT 400 CENTRES

175×38 RIDGE

A

175, 225 OR 250×75 PURLINS

350×50 RUNNERS

100×50 COLLARS
100×50 HANGERS
100×50 BRACES
AT EVERY FOURTH
SPAR – SEE "F"

100×50 COLLARS
AT 400 CENTRES

100×75 HEAD

100×50 STUDS

100×38 NOGGINGS

WINDOW

LATH &
PLASTER

WINDOW

2590

C

175×19 FASCIA

G

DORMER
WINDOW

PLASTER
100×50 WALL PLATE
28 T & G. BOARDING
225×50 JOISTS
100×20 SKIRTING

STOOTHED PARTITION

BEDROOM FLOOR

DOOR

55°

CUPBOARD

B

275

275

7315

SECTION THROUGH ROOF ETC.

SCALE FOR A

0   1000   2000   3000   4000   5000
MM

FELT

38×19 BATTENS
PLAIN TILES
125 DIA. HALF-
ROUND C.I.
GUTTER

175×19 FASCIA   275

D

100×50 SPARS
100×50 WALL PLATE
225×75 PURLIN
BEAMFILLING
ARCHITRAVE
100×38 CASING
FLUSH DOOR

E

19 BEAD

100×50 SPARS AT 400 CENTRES
100×50 COLLARS AT
EVERY FOURTH SPAR

250×75
PURLIN

DETAILS 'B' & "C"

SCALE FOR DETAILS

0   500   1000
MM

100×50 BRACES AT EVERY FOURTH SPAR

100×50 COLLARS AT 400 CENTRES

100×50 RUNNER

100×75 HEAD
& 100×50 STUDS
OF PARTITION

100×50 RUNNER

100×50 HANGERS

F

SKETCH SHOWING
ROOF CONSTRUCTION

Fig. 94.

Alternatively, the hangers may be placed with their edges adjacent to and notched over the purlins and runners, and nailed to the spars, etc. It will be observed that, like the upper collars, the braces and hangers are spaced at every third, fourth or fifth spar apart.

Either double floor joists (as shown at A) or a single joist of sufficient thickness should be provided to support the central partition.

The sketch at F, Fig. 94, shows more clearly much of the construction described above.

The shape of the room is improved if, as illustrated at C, a partition is fixed along each side below the lower set of purlins ; the space between each outer wall and partition may be utilized for storage ; the head of the door is detailed at E. An alternative arrangement is shown at G, where a dormer window is indicated in broken outline ; if required, a partition with returned ends at the window may be provided (see broken vertical line continued from the purlin).

The closed eaves detail at D shows a fascia board which is nailed to the square cut feet of the spars and a triangular fillet secured to the backs of the spars. The lower edge of the fascia is scribed (p. 141) to the wall. Fig. 102 shows an alternative closed eaves.

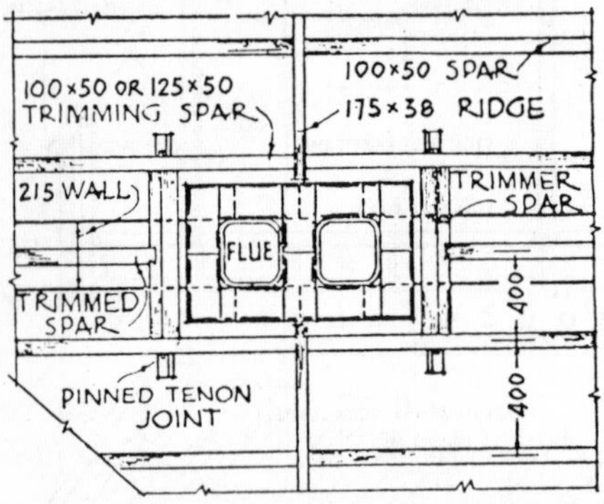

TRIMMING.—This is required at chimney stacks, dormer windows, skylights, etc. and the construction is very similar to that for floors (p. 93). The names of the various spars are also similar to those applied to floor trimming, *i.e.*, *trimming spars* (or rafters), *trimmer spars* and *trimmed spars*.

PLAN OF TRIMMING
TO CHIMNEY STACK
Fig. 95.

The trimming of a roof round a chimney stack which penetrates at the ridge is shown in Figs. 92, 93 and 95. The plan of the stack in Fig. 95 shows the various members. The joint between the trimmer and trimming spars may be either a tusk tenon (see Fig. 51) or a similar joint without the tusk, called a *pinned tenon joint*. That between the trimmed and trimmer spars should be either a dovetailed housed joint (Fig. 52) or a bevelled haunched joint described on p. 95. In cheap work the trimmed spars are simply butt-jointed and nailed to the trimmers.

The trimming round a chimney stack which penetrates a roof some distance between the eaves and ridge is shown in Fig. 96. The construction should be clear from the foregoing description. The gutter at the back of the stack and shown at E and F is referred to on p. 165. The slating and leadwork have been included in these two sections in order to give

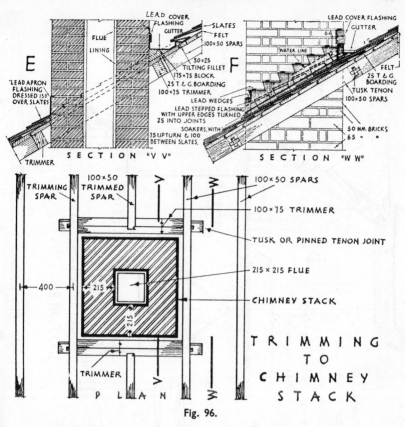

Fig. 96.

some indication of the work in these trades which is necessary to make the roof watertight.

DORMER WINDOWS.—These can be used to provide daylighting within the roof space and are often found in bungalows to provide habitable rooms within the roof structure. When rooms are made in this position it should be remembered that the ceiling height must be at least 2·3 m over not less than half the area of the room measured at a height of 1·5 m above the floor level ; (a Building Regulation requirement). For the best appearance a dormer window should be made to fit well down on the roof so that there is the minimum upstand from window sill to the main roof, see B, Fig. 97. The dormer roof may be flat or pitched.

In the example in Fig. 97 of a pitched dormer the roof is covered with plain tiles to match the main roof and the junction between the dormer roof and main roof is formed with a swept valley.[1] A half elevation and

---

[1] In a swept tile valley the junction between the two roof planes is blocked out with a valley board about 300 mm wide. The tiler can then cut the tiles and " sweep " the fixed tiles round in a gentle curve thus eliminating the valley gutter (compare this with the lead valleys in Fig. 119).

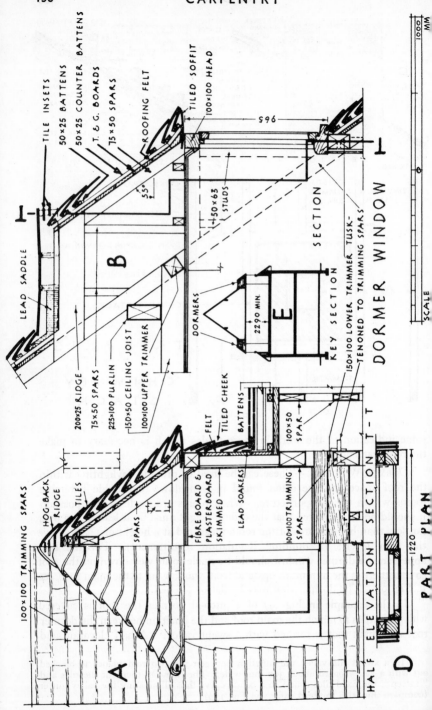

DORMER WINDOW

TILE INSETS

50×25 BATTENS

50×25 COUNTER BATTENS

T. & G. BOARDS

75×50 SPARS

ROOFING FELT

TILED SOFFIT

100×100 HEAD

965

55°

50×63 STUDS

LEAD SADDLE

B

200×25 RIDGE

75×50 SPARS

225×100 PURLIN

150×50 CEILING JOIST

100×100 UPPER TRIMMER

DORMERS

SECTION

KEY SECTION

E

2290 MIN.

150×100 LOWER TRIMMER TUSK-TENONED TO TRIMMING SPARS

SECTION

100×100 TRIMMING SPARS

HOG-BACK RIDGE

TILES

SPARS

FELT

TILED CHEEK

BATTENS

FIBRE BOARD & PLASTERBOARD SKIMMED

LEAD SOAKERS

100×100 TRIMMING SPAR

100×50 SPAR

A

HALF ELEVATION

SECTION T-T

D

PART PLAN

1220

SCALE

1000 MM

section are drawn at A with another section at B. To form the necessary opening in the roof 100 mm by 100 mm trimming spars are placed on either side and to these are joined an upper 100 mm by 100 mm and a lower 150 mm by 100 mm trimmer spars. These joints may be of the bevelled housed type or tusk tenoned (for half the thickness only in the case of the lower trimmer, as shown). A window frame forming the front of the dormer and having 100 mm by 100 mm sides and head with 200 mm by 100 mm sill rests on the lower trimmer which has 75 mm by 50 mm studs beneath it for the internal lining of fibreboard covered with plasterboard. The sides (or cheeks) of the dormer have a 100 mm by 100 mm head joined to that of the window frame and upper trimming spar, 50 mm by 63 mm studs at the cheeks are covered externally with boarding, felt and nibless vertical tile hanging. Concealed lead soakers are used at the junction of the cheeks and main roof. The dormer roof is covered with tiles in the usual way. The top tiles beneath the window sill are bedded on a lead apron which is tucked into a groove in the centre of the underside of the sill. The apron is scored and triangular pieces are cut out of it for the mortar bedding; the joint between sill and top tile is then mastic pointed. The eaves detail is different to those described below in having its soffit formed with plain tiles turned upside down. As the roof is small a gutter would be clumsy and unnecessary.

EAVES DETAILS.—The various types of eaves are defined on pp. 113 and 114. They are :—(1) flush, (2) open, (3) closed and (4) sprocketed eaves. Careful attention should be paid to the design of the eaves. Over-elaboration should be avoided, the simpler the detail the better. It is a common mistake to use an excessively deep fascia, and the clumsy effect which this produces is shown in Fig. 98. Occasionally sprockets are used with an inadequate pitch (pp. 113 and 142) with the result that rain and fine snow are driven up between the slates, etc. to cause dampness.

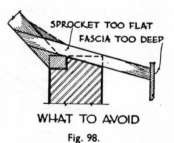

WHAT TO AVOID

Fig. 98.

1. *Flush Eaves.*—An example of this type, suitable for a flat roof, is detailed at E, Fig. 73. The fascia is only sufficiently deep to cover the ends of the joists to which it is nailed or screwed; the lower edge is occasionally moulded. The eaves gutter (of cast iron or asbestos-cement) is supported by wrought iron brackets which are twice screwed (by the plumber) to the fascia. The top edge of the outer board should be rounded off to prevent damage to the roof covering.

Another example is shown in Fig. 99. It is assumed to be the detail of the eaves at E, Fig. 81, although it may be applied to any of the other pitched roofs. The fascia is secured to the feet of the spars (birdsmouthed and nailed to the wall plate) and it projects slightly above the boarding in order to give the necessary tilt to the slates, shown by broken lines (see also Fig. 81). Roofing felt is shown by broken lines; this is nailed to the boarding by the slater. If slating battens are used in lieu of the t. and g. boarding, the upper portion of the detail will resemble that in Fig. 102.

The details at B and E, Fig. 114, also show a flush eaves.

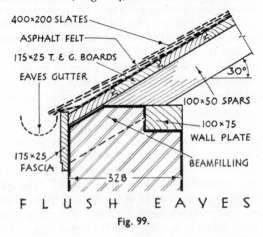

FLUSH EAVES
Fig. 99.

2. *Open Projecting Eaves.*—This is shown in Fig. 100. It is the detail of the eaves at K, Fig. 77. The feet of the spars project 150 mm and are shaped as shown or as indicated in Fig. 101. It is not necessary to provide a fascia to an open eaves ; the straps supporting the gutter are either

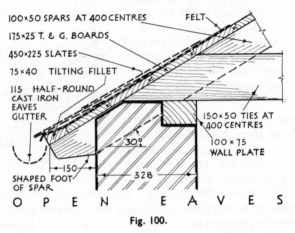

OPEN EAVES
Fig. 100.

fixed to the boarding or the sides of the spars, and the required tilt to the slates is obtained by the tilting fillet. The roofing felt and slates are indicated by broken lines.

Another simple eaves of this type is detailed in Fig. 101. Plain tiles are shown as the covering material hung on battens ; normally, the tiles in every fourth course only are nailed.

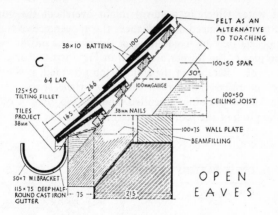

FELT AS AN
ALTERNATIVE
TO TORCHING

38×10 BATTENS

C

100

100×50 SPAR

50°

64 LAP

266

100mmGAUGE

125×50
TILTING FILLET

100×50
CEILING JOIST

165

TILES
PROJECT
38mm

38mm NAILS

100×75 WALL PLATE

BEAMFILLING

50×7 W.I.BRACKET

OPEN
EAVES

115×75 DEEP HALF-
ROUND CAST IRON
GUTTER

75

215

Fig. 101.

3. *Closed Projecting Eaves.*—There are two forms of closed eaves, *i.e.*, those with sprockets and those without.

An example of the latter is shown in Fig. 102. This is a detail of the eaves at D, Fig. 84. The ends of the spars are sawn to the shape as shown, the soffit board is nailed to the spars, after which the fascia is secured to the spars, and nails are also carefully driven through the fascia into the edge of the soffit board. In first class work, a continuous groove, about

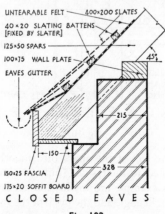

UNTEARABLE FELT

400×200 SLATES

40×20 SLATING BATTENS
[FIXED BY SLATER]

125×50 SPARS

100×75 WALL PLATE

45°

EAVES GUTTER

215

150

328

150×25 FASCIA

175×20 SOFFIT BOARD

CLOSED EAVES

Fig. 102.

10 mm deep is formed along the back of the fascia to receive the edge of the soffit board ; no unsightly gap then appears between the two boards if the soffit board shrinks. For greater projections, soffit boards nailed to soffit bearers (see next page) are employed.

A closed eaves is shown at D, Fig. 94, and described on p. 134.

4. *Sprocketed Eaves.*—Sprockets (p. 114) are usually fixed at the eaves of steeply pitched roofs. They reduce the pitch of the roof, and so lessen

the flow of water which in a storm might overshoot the gutter. The bell-shaped finish produced also enhances the appearance of the roof.

The sprockets may be fixed either at the sides or on the backs of the spars. Examples of the former are given in Figs. 103 and 105, and of the latter in Fig. 106. Because of the solid fixing provided, it is easier to nail sprockets to the backs rather than the sides of the spars. Accordingly, the detail in Fig. 103 is often amended, and the top ends of the sprockets are cut to the pitch of the roof (as AB) and nailed to the backs of the spars.

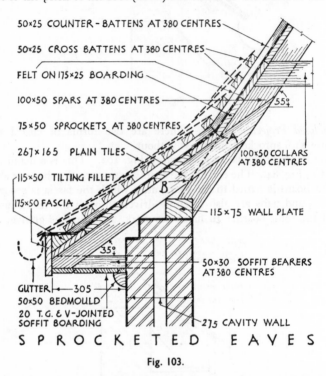

50×25 COUNTER-BATTENS AT 380 CENTRES

50×25 CROSS BATTENS AT 380 CENTRES

FELT ON 175×25 BOARDING

100×50 SPARS AT 380 CENTRES

75×50 SPROCKETS AT 380 CENTRES

267×165 PLAIN TILES

115×50 TILTING FILLET

175×50 FASCIA

55°

100×50 COLLARS AT 380 CENTRES

115×75 WALL PLATE

35°

GUTTER 305

50×50 BEDMOULD

20 T.G. & V-JOINTED SOFFIT BOARDING

50×30 SOFFIT BEARERS AT 380 CENTRES

275 CAVITY WALL

S P R O C K E T E D   E A V E S

Fig. 103.

Fig. 103 shows the feet of the spars birdsmouthed to the wall plate, and the sprockets nailed to their sides (but see above) and to the plate. The sprockets in this example are inclined at an angle which equals the difference between a right angle and the pitch of the roof, *i.e.* $90° - 55° = 35°$; this produces a graceful sweep. A fascia board is fixed to the ends, and a tilting fillet nailed to the boarding. The underside is enclosed with soffit boarding nailed to *soffit bearers* secured at the outer ends to the spars with their inner ends supported on the wall and in pockets left by the bricklayer; these latter ends are either built in by the bricklayer or wedged tight between the timber and the sides of the pockets. As indicated, this boarding consists of *tongued, grooved and vee-jointed boards*; an enlarged detail of these is given in Fig. 104; the ends of the boards should be carefully mitred at hipped ends, etc. (see s, Fig. 90); the outer

board may be housed into the fascia, as described on p. 139. The small
bedmould (quadrant shaped) covers any gap which may be caused if the
boarding shrinks; this is *scribed* and plugged to the wall. " Scribe "
means to mark for accurate fitting, and in this case scribing is necessary

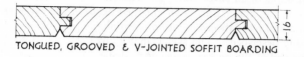

TONGUED, GROOVED & V-JOINTED SOFFIT BOARDING

Fig. 104.

to ensure that the back of the mould fits the relatively irregular surface of
the brickwork. Scribing is done with the aid of compasses (see 5, Fig. 155);
the mould is placed slightly away from but parallel to the wall, and with
the points slightly apart, the compasses are drawn horizontally along it,
with one point contacting and following the irregularities of the wall and
the other marking a parallel line on the mould; the back of the mould
is sawn to this irregular line, and thus the fixed mould will fit tightly
against the brickwork.

In this example (Fig. 103) the butt jointed roof boarding is shown
covered with felt, fixed by the tiler; counter-battens (indicated lightly)
running from eaves to ridge at 400 mm centres are then nailed to receive the
cross tiling battens. This construction is only adopted in best work. Plain
tiles (see Fig. 101) form the covering material.

The two eaves in Fig. 105 have side sprockets. With exception of the
larger bedmould and the difference in the groundwork for the covering
material, that at L resembles Fig. 103. The detail at K is also similar,
except that the wall ends of the soffit bearers are fixed to a fillet which
is plugged to the wall. The sprockets are shown on the plan F, Fig. 90;
those nailed at the sides of the hip rafters are necessary to provide a means

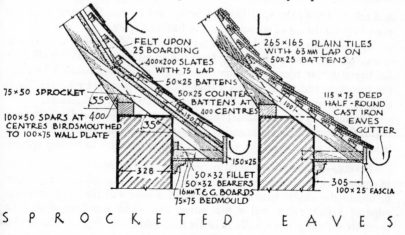

S P R O C K E T E D   E A V E S

Fig. 105.

of fixing the upper ends of the short sprockets at the corners and the bearers to which the fascia (mitred at the angles) and the mitred ends of the soffit boards are nailed ; one of these bearers is shown at T, Fig. 90, but has been omitted at S in order to show the mitre between the soffit boards.

An example of an eaves with sprockets fixed on the backs of the spars is shown in Fig. 106. This is an enlarged detail of the eaves of the roof in Figs. 92 and 93. Its construction resembles that already described. The sketch in Fig. 93 shows one end of the nearest spar cut to shape, the next spar is shown with the sprocket fixed,

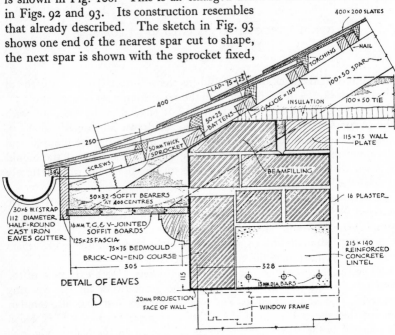

Fig. 106.

and the next with the sprocket and bearer fixed. See also note on p. 140 regarding the fixing of sprockets to the backs of spars.

Sprockets should not be given an inadequate slope such as is shown in Fig. 98, for, besides detracting from the appearance, it makes it difficult for the slater or tiler to negotiate the angle at the intersection between the sprockets and spars, unless a triangular fillet (shown by broken lines) is fixed. A flat slope is also difficult to make watertight at the eaves.

The torching (mortar fillets) shown between the slates at the battens in Fig. 106 is done by the slater and is necessary to prevent rain and snow from being blown in between the slates. It is an alternative to roofing felting.

BEVELS.—The ends of certain roof members must be accurately cut to ensure a tight fit between adjacent timbers. For example, it is necessary to obtain the angles between the head of a spar and the ridge, the top edge of a jack rafter and a hip rafter, etc. The members should be carefully set out geometrically in order to obtain these angles. The angles are known as *bevels* (a bevel means " an inclined direction on the face, edge, or both, of a member giving the line for cutting ") or *cuts*, and the tool called the *bevel* (see 3, Fig. 155) is used for applying the angles direct to the wood members to give the required cutting lines. It is only possible here to describe briefly how some of these bevels are obtained.

The following description applies to the several members shown in outline in the elevation and part plan of the hipped end of a double roof shown in Fig. 108 ; it is assumed that the roof is rectangular on plan.

Spar or Common Rafter Bevels.—These are shown at A, Fig. 107. The spar is drawn at the required pitch (30° in this case), and the angle at the wall plate (called the *foot* or *plate bevel*) and that at the ridge (known as the *head* or *plumb bevel*) are obtained direct. The true length of the spars is obtained from the elevation.[1]

The bevel tool (3, Fig. 155 and p. 203) is applied to the drawing, its slotted blade is adjusted and set to the head bevel which is then transferred

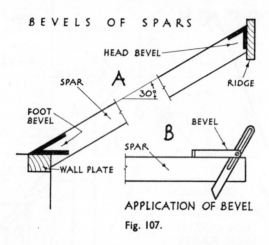

BEVELS OF SPARS

HEAD BEVEL

SPAR

A

30°

RIDGE

FOOT BEVEL

B

BEVEL

SPAR

WALL PLATE

APPLICATION OF BEVEL

Fig. 107.

to a spar ; the foot bevel is obtained and transferred in a similar manner. The application of the head bevel at the end of a spar is shown at B, Fig. 107. A pencilled line is drawn on the side of the spar along the edge of the blade. The back of the spar is " square-marked," *i.e.*, the try square (see 2, Fig. 155) is applied to the top edge of the spar where it is intersected by the above line, and a line is drawn on the back of the spar along the edge of the blade. The bevel is then adjusted to the angle at the foot of the spar and applied at the opposite end in a similar manner and at the correct distance along the top edge ; the edge is square-marked. When the foot is birdsmouthed as shown at A, Fig. 107, the try square is used against the bevel to give the required vertical mark ; the bottom edge is square-marked. Both ends are now shaped by sawing along the lines. This is called a *pattern rafter*, as it is applied as a pattern to all the spars,

---

[1] The clear span of the roof should be checked on the job.

i.e., it is placed in turn on the side of each spar and the latter is marked to conform to the shaped ends of the pattern.

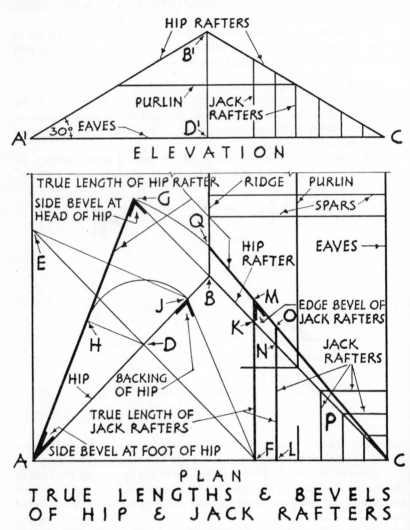

Fig. 108.

Hip Rafter Bevels (see Fig. 108).—The hip rafters in plan are shown at AB and BC, and A′B′ and B′C′ in elevation.

From B draw BG at right angles to AB and of length equal to height B′D′. Draw AG; this is the true length of the hip rafter. Angle BAG is the *side* bevel at the foot of the hip rafter and angle AGB is the *side* bevel at the head of this rafter.

It is occasionally necessary (as in good class work when the spars are covered with boarding) to shape the *top edge* of a hip rafter to the angle between that of the adjacent roof surfaces. This is known as the *dihedral angle*, and the shaped edge of the rafter is called the *backing* (see c, Fig. 109). The following is the geometrical construction for obtaining the angle of backing as shown in the plan in Fig. 108. From any point D along AB draw EF at right angles to AB. From D set up DH at right angles to AG. With D as centre and DH as radius, describe arc HJ Draw EJ and JF. The angle EJF is the dihedral angle or angle of backing.

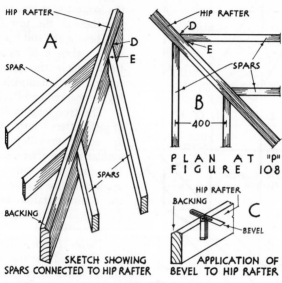

Fig. 109.

The angle at which the bevel (3, Fig. 155) is set is DJF, as indicated. The application of the bevel is shown at c, Fig. 109.

The length and side bevels of valley rafters (see A, Fig. 72) are obtained in a similar manner. Unlike hip rafters, however, the top edges of valley rafters are not required to be bevelled (see Fig. 119).

Jack Rafter Bevels.—The centre lines of some jack rafters have been set off on the plan in Fig. 108.

The true lengths of jack rafters are obtained in the following manner : With centre A and radius AG describe arc GQ. Join QC and continue the lines representing the rafters on plan until they meet this line QC. The true length of each rafter is the distance measured from the eaves to QC. Thus, the true length of jack rafter FK is FM and that of LN is LO.

A portion of this plan showing the thickness of the members is given at B, Fig. 109, and an axonometric sketch of it appears at A. These indicate that the top edges of the jack rafters are bevelled against the hip rafter—see DE. The *edge* or *cross bevel* is obtained from the development on Fig. 108 and is angle FMC. This bevel is common to all the jack rafters when the roof surfaces are equally inclined.

As indicated at A, Fig. 109, the *side* cuts are vertical (called the *down* or *plumb cuts*), and the bevel is therefore the same as that for the head of the spars. The foot bevel is also that of the common rafters (see A, Fig. 107).

It is usual to mark the true lengths of all the jack rafters on the longest spar and after marking thereon the top edge and side bevels, the end is cut. This spar is then used as a pattern from which the shorter jack rafters (which diminish uniformly) are marked and cut in pairs.

The true lengths and bevels of jack rafters whose lower ends are nailed to valley rafters are similar to the above, except that the edge and side cuts are applied at the lower ends, and the common rafter head bevel (A, Fig. 107) is applied at the upper ends.

Purlin Bevels.—Two bevels are required at the end of a purlin when it is nailed to the side of a hip rafter, and when the purlin is fixed at right angles to the pitch of the roof.

One development adopted to obtain these bevels is shown in Fig. 110. Draw the line of pitch of the roof and a section through the purlin at right angles to the pitch. Below this draw portions of the hip rafter at 45° and the purlin abutting against it. From A, B and C drop vertical lines to D, E and F. Draw a horizontal line at B, and with B as centre describe arcs AG and CH. From G draw a vertical line until it meets at J, the short horizontal line from D. Similarly, drop a vertical line from H and a horizontal line from F, meeting at K. Draw EJ and EK. The *edge* or *cross bevel* is angle GJE, and the *side* or *down bevel* of the purlin is angle HKE.

The bevels of purlins which have their ends nailed to valley rafters are determined in a similar manner.

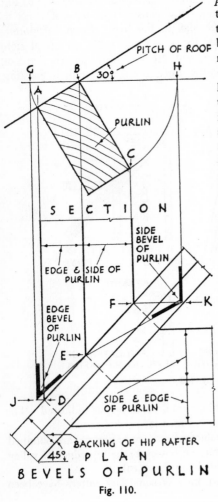

Fig. 110.

When the purlins are vertical (as shown in Fig. 93), their ends are only required to be cut to a 45° mitre, *i.e.*, the plan angle of the hip or valley rafters.

NOTE.—The aforementioned bevels are usually developed on the job. A suitable scale for the plan (such as that in Fig. 108) is 1 : 10 ; a larger scale, say one-fifth full size, may be adopted for developing the purlin bevels. All the bevels are taken from the drawing and applied to the members, as explained on p. 143. With experience, coupled with a sound knowledge of geometry, the apprentice will be able to adopt certain shortcut methods for developing bevels.

ERECTION.—The spars, cut to length and with their ends shaped to the required bevels, are fixed in the following manner : If the roof is of the simple type shown in Fig. 81, it is usual first to fix the two pairs of end spars (called the *pitch spars*) to the ridge and wall plates, followed by the remainder. Generally, the ridge must be level, and as a temporary support until the end spars are fixed, it may be carried at each end and at the correct height by a vertical strut ; this has its upper end notched as shown in Fig. 111 to receive the ridge. The intermediate spars are fixed after the end pairs have been secured. *It is most essential that the wall plates, to which the feet of the spars are fixed, are level and parallel.* If this is observed, and the spars are cut to the correct length, many errors in fixing are obviated and there will be no difficulty in determining the position and level of the ridge. For a long ridge, an intermediate vertical strut is required to prevent the ridge sagging.

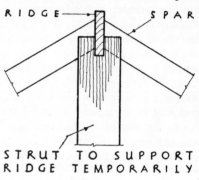

Fig. 111.

If it is a gabled roof (see A, Fig. 72), the ridge is often supported on a corbel of one or two projecting bricks (resembling Fig. 78) built at the correct level. The bricklayer should use a pair of pitch spars to ensure that each corbel is built at the correct position with its upper surface level with the underside of the ridge.

If the ridge is to be intercepted by a chimney stack, sometimes the work on the stack is suspended when it has reached a height just below the position of the ridge. The ridge and spars (excepting the trimming and trimmed spars) are then fixed, a portion of the ridge (equal to the width of the stack) is cut and removed, and the bricklayer then completes the stack. Alternatively, a corbel is built at each side of the stack to support the ridge. The trimming, trimmer and trimmed spars are then fixed (p. 134).

When the roof is of the type shown in either Fig. 83 or 84, the ties or collars are, of course, fixed after the spars have been secured to the wall plates and ridge. Any notching of the spars (such as is required if the ends of the ties or collars are dovetail halved jointed—see Fig. 85) should be done before the spars are erected. If the work has been correctly set out (with the wall plates level and parallel, etc.) much of this work is done at the " shop ".

If the spars project and their lower ends are to be shaped, as in Fig. 100, it is usual to do this at the workshop. Thus, the ends are cut to shape, planed smooth (with exception of the backs) and primed (given a first coat of paint) before the spars are sent on to the job. The end spars and ridge are fixed as described above, followed by the intermediate spars. The spars should have been cut slightly in excess of the finished length in order that their heads may be cut and fitted to the ridge. When fixing these intermediate spars, care should be taken to ensure that their feet are in alignment with a line stretched between the feet of the end spars. If there is any inaccuracy in the brickwork, it may be advisable to reverse the above order. Thus, in turn each intermediate spar, head cut to the required bevel, is placed temporarily in position with its head pressed against the ridge until the position of the square cut at the foot is marked when in alignment with a line stretched between the two end feet ; the wall line is marked on the lower edge, the spar is removed, and after its feet have been shaped is returned and finally fixed.

In the case of a sprocketed eaves, such as that in Fig. 103, the spars are first secured, followed by the two end sprockets (and an intermediate sprocket if the eaves is a long one). The fascia is temporarily nailed to these sprockets. The intermediate sprockets, having been shaped (preferably at the shop before being sent on to the job) are then fixed with their projecting ends in contact with the fascia. The latter is removed to permit of the convenient fixing of the soffit bearers ; these bearers, after being fixed to the spars, are wedged in the wall pockets (p. 140). The fascia is then nailed and fixed permanently, followed by the soffit boarding and bedmould, the latter being scribed and plugged to the wall (p. 141). All of this work at the eaves is done on the scaffold left by the bricklayer.

The members of a double roof (such as that shown in Fig. 86) are generally fixed in the following order : First, the wall plates are checked for " dead level and parallelism " and then the ridge with the two end pairs of spars are secured. The purlins (having been lifted approximately in their position and used as a scaffold to stand on) are then placed in correct position, their ends being firmly wedged in pockets in the gable walls (or supported on corbels if they are party walls—see p. 101) and subsequently built in. If the spars are to be cogged to the purlins (Fig. 87), the cogs are formed on the purlins before fixing, and the end spars should be notched as required. The remaining spars are notched at the proper level and nailed to the ridge, wall plates and purlins. If the roof has hipped ends, the hip rafters are fixed to the ridge and dragon beams (previously secured to the angle ties—see Fig. 91), followed by the purlins and remaining rafters.

TRUSSED RAFTER ROOFS

This type of roof (Fig. 112) consists of planed stress graded timbers fastened together in the form of trusses and placed at relatively close distances apart. It comprises rafters (spars) joined to ceiling joists and intermediate members. The trusses are prefabricated and because roofs of this type are more quickly erected and use much less timber than purlin roofs they have almost entirely replaced them for housing work.

The trusses are placed at centres not exceeding 600 mm [50 mm by 25 mm (min.) tiling battens must be used] and the separate members of the truss, which must be of the same thickness, are joined by plywood gussets glued and nailed to *each* face at the joints. The adhesive used for this purpose is commonly resorcinol/formaldehyde (see Chap. XI); 40 mm 12 gauge galvanized nails at 100 mm centres in two rows per member are used (see c). Alternatives to plywood gussets are 18 or 20 gauge galvanized metal plates which may be either perforated for fastening with clout nails or may have integral teeth in which case a special press is needed to make the joint.

The two most popular truss shapes are the W or Fink type (see A and D) or the Fan type (B) for larger spans. Notching or birds-mouthing of the truss should not be allowed. The moisture content of the timber should be 22 per cent. or less so care should be taken to fix roof coverings quickly; trusses should be stacked flat on a level base before erection.

The following Table gives the sizes of the timbers in mm and spans in m for the two types of truss.

TABLE VI. Maximum Spans (m) for Fink and Fan Trussed Rafter Roofs

Basic* size of member (mm) (*Actual size 3 mm less to allow for planing)	Pitch (degrees)				
	15	20	25	30	35
38 × 100	7	7·7	8·1	8·3	8·6
38 × 125	8·7	9·2	9·5	9·8	10·1
44 × 75	5·5	6	6·5	6·9	7·3
44 × 100	7·5	8·2	8·7	8·9	9·2
50 × 75	5·9	6·5	6·9	7·4	7·8
50 × 100	7·9	8·7	9·2	9·5	9·8

TRIPLE ROOF

As stated on p. 126, the unsupported length of purlins should not exceed 4·8 m, otherwise their size will be unduly large and uneconomical. Therefore, roof trusses (sometimes called *principals*) must be provided if cross division walls are not available as supports to limit the purlins' length. Accordingly, and as implied by its name, a triple roof consists of three sets of members, *i.e.*, *spars* that distribute the weight of the roof covering, snow and wind pressure to the *purlins* which transmit this load to the *trusses*, and these in turn transfer the weight to the walls. As the members comprising the truss are framed together, this is also known as a *framed roof*. The outline of a truss must conform to the shape of the roof.

There are two forms of wood truss principals, *i.e.* (1) the built-up truss, (2) the glued laminated portal frame. An older type of truss, once widely used, was the king post truss ; it has been replaced by (1) above and is shown here for reference purposes because it is often found in alteration work.

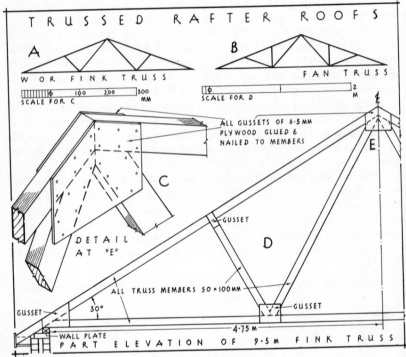

Fig. 112.

1. BUILT-UP ROOF TRUSS.—This, the most popular timber truss, is so called as the principal rafters, main tie and struts are double members with a space between equal to the thickness of the inclined ties. It has replaced and is more economical than the king post and queen post roof trusses, especially in the labours involved at the joints, and can be used for spans up to 14 m. The design and size of members depend upon the span, distance apart of the trusses, quality or grade of the timber employed and the weight of the covering material.

The example illustrated by a part elevation in Fig. 113 is typical for a clear span not exceeding 9 m, truss-spaced at about 3·7 m centres and supporting a light roof covering, such as asbestos-cement corrugated sheets which are up to 3 m long and 1 m wide. The roof has a pitch of one-quarter (26½°—see p. 113). Note that the principals, main tie (or tie beam) and struts are double members and that the inclined ties are single members passing between the principals and main tie.

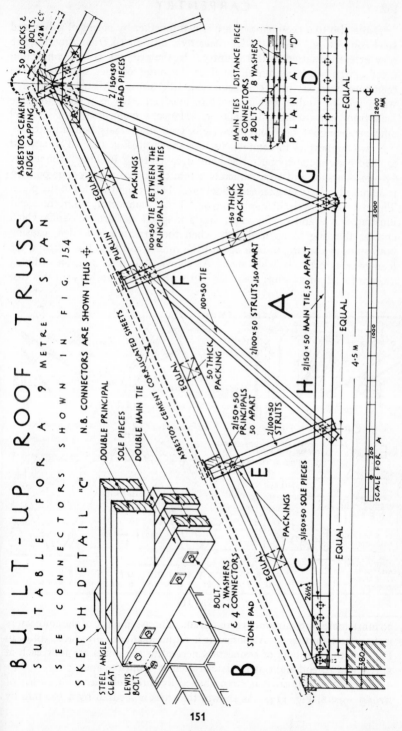

BUILT-UP ROOF TRUSS

SUITABLE FOR A 9 METRE SPAN

SEE CONNECTORS SHOWN IN FIG. 154

N.B. CONNECTORS ARE SHOWN THUS ✛

SKETCH DETAIL "C"

50 BLOCKS &
9 BOLTS,
1/2M C/C

ASBESTOS-CEMENT
RIDGE CAPPING

2/150×50
HEAD PIECES

MAIN TIES   DISTANCE PIECE
8 CONNECTORS   8 WASHERS
4 BOLTS

PLAN AT "D"

PACKINGS

100×50 TIE BETWEEN THE
PRINCIPALS & MAIN TIES

150 THICK PACKING

G

PURLIN

EQUAL

F

100×50 TIE

2/100×50 STRUTS, 150 APART

A

50 THICK PACKING

DOUBLE PRINCIPAL

SOLE PIECES

DOUBLE MAIN TIE

ASBESTOS-CEMENT CORRUGATED SHEETS

EQUAL

2/150×50 PRINCIPALS
50 APART

2/100×50 STRUTS

H   2/150×50 MAIN TIE, 50 APART

EQUAL   4·5 M   EQUAL

STEEL ANGLE
CLEAT

LEWIS
BOLT

BOLT,
2 WASHERS
& 4 CONNECTORS

STONE PAD

B

PACKINGS

3/150×50 SOLE PIECES

E

C   26½?

380·?

EQUAL

SCALE FOR A

200   1000   2000   2800 MM

Fig. 113.

151

Note also that the members are joined together by means of galvanized steel connectors. These *timber connectors*, introduced from America, are now extensively used in this country. In the ordinary bolted connection most of the stress is concentrated at the outer contact surfaces of the members, whereas in a connector joint the stress is distributed over a wider area of timber and is therefore more effective in transmitting the load from one member to another. Hence a connector joint is more efficient, being stronger and more rigid, than an ordinary bolted joint.

Two connectors in common use are shown in Fig. 154. The circular toothed plate metal connector at A and C, called the " Bulldog " connector, 63 mm diameter and approximately 1 mm thick, has twenty-four pointed triangular teeth (twelve on each side) turned off at right-angles to the plate. The metal 63 mm circular split ring or " Teco " connector, 3 mm thick and illustrated at B and C, Fig. 154, has a tongued and grooved split which enables the ring to open slightly when the structure is loaded, causing both sides of the ring to bear against the timber and thus transmitting the load over a relatively wide area. As shown at C, Fig. 154, an ordinary 12 mm (or 16 or 20 mm) diameter bolt with nut and washers (in this case

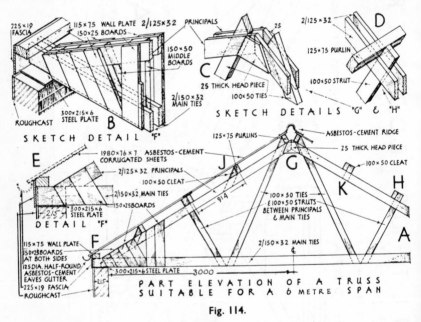

Fig. 114.

50 mm square by 8 mm thick) is required. The assembling of the connectors is described on p. 201. A connector is required at each interface; thus, in Fig. 113 the numbers of connectors needed to the joints are four each at E, F, G and H, and eight at D, sixteen each at C and the apex.

The foot of the truss is strengthened by three *sole pieces* (or *gussets* or *straps*)—see B, Fig. 113. As shown, the truss is secured by a 100 mm by

100 mm by 6 mm or 10 mm steel angle cleat (one at each side) with a 13 mm diameter bolt through the main tie (and connectors) and a 16 mm by 100 mm long *lewis bolt* (or *rag bolt*) let into the stone pad.

The principals at the apex are secured by a *head piece* (or *gusset*) at each side, as shown. The purlins are bolted to the projecting ends of the struts (see also D, Fig. 114) ; those at the apex are either bolted or spiked to the ends of the ties and, as indicated, shaped wood blocks at 1·2 m intervals are bolted to the purlins as a further safeguard against the latter tilting. Purlins should be spliced over trusses.

Long members, such as main ties, may be in two half-lengths and spliced together with a packing between—see D, Fig. 113. This also enables a truss to be transported to the job from the workshop in two halves which are joined together on the site. Trusses for large spans may have two splices at one-third points in the main ties, and one splice at mid-point of each principal ; gussets (as a. B) may be used here as stiffeners.

Principals and long struts are stiffened by the provision of packings or distance pieces, as shown, of thickness equal to the distance between adjacent members, and securely nailed. Note that these packings are required also at E and the apex (for the two outer sets of connectors only).

It will be observed that this is not a triple roof, as spars are not provided (see pp. 115 and 149). Spars are not necessary if large sheets, such as the corrugated sheets shown, are adopted as these are secured by driving screws at the purlins and wall plates. If the covering material was of small and heavier units (such as tiles or slates) spars would be needed, the pitch would have to be increased and larger sizes of the main ties and principals at least would be necessary.

Another built-up type of truss, shown in Fig. 114 and called a *laminated wood truss* as it is composed of relatively thin members, is employed when the covering material is light in weight. It has been extensively adopted for semi-permanent buildings, farm buildings, etc. The members are simply well spiked at the joints. The trusses are spaced up to 300 cm apart. Intermediate purlins are nailed to cleats, as at J or K.

2. GLUED LAMINATED PORTAL FRAME.

—A feature of laminated construction is that relatively thin pieces of timber are glued together to build up the section required. The laminates may be placed vertically or horizontally (see H, Fig. 115) and in the former case, the bending stresses can be increased by approximately fifty per cent., depending on their number. This increase is permissible because the timber beam is " man made " and defects such as knots (which reduce the strength) can be eliminated by careful selection, or in any case allowed to extend no further than one laminate thickness. Frequently, the joint between the thin scantlings is horizontal (as in the roof part of the portal in Fig. 115), for in this way curved or arched members are readily fabricated.

In addition to the ease of bending, laminated members have the following advantages. As the parts are thin, they can be seasoned more quickly than large baulk stuff, and use can be made of short lengths which might otherwise be wasted. For horizontally laminated beams, good quality timber can be used for the highly stressed parts (the top and bottom extreme fibres), with second-grade wood towards the centre of the section. The use of scarf joints (see F') enables long units to be built up ; these joints are usually made with a slope of 1 : 12 (steeper for thin laminates) and they are spaced not less than 1·2 m apart so that they do not all occur at the same section.

Manufacture of Laminated Portal.—After the scarf joints have been glued together, the laminates are placed in a shaped jig, each interface being spread with glue. Depending on the thickness of the member and its curvature, up to twenty laminates can be glued and positioned at a time, working from the innermost one and finishing with the outer ; clamps are used to ease them against the bend of the jig.

There is a wide choice of adhesives, casein or resorcinol resin glue being commonly adopted. Most of the work is done with the former but the latter should be used in wet or humid conditions ; they should be used in accordance with the instructions of the maker (see pp.225-228).

Laminated timber is of course more costly than the solid wood but unlike the latter, large sections can be shaped by bending. The radius of the curve should not normally be less than 150 times the laminate thickness (approximately 22 mm thick and planed) or 100 times where specially selected straight-grained wood is used.

The example in Fig. 115 gives the arrangement in the key diagram at D, where the portal is in two halves jointed at the apex A. At this point it is 200 mm deep (see E) and 190 mm wide (200 mm nominal) ; two 12 mm dia. m.s. locating pins help to secure the joint. The fixing also includes a handrail-type bolt where the nuts are tightened through slots from above.

The detail at G shows the base of the portal fitting into a shoe formed of 10 mm thick m.s. having four holes for the holding-down bolts. A plate halfway up the shoe coincides with the floor level, the space beneath being filled with concrete. The portal is secured to the shoe with a bolt and connector[1] plates, the latter between the timber and the shoe.

The detail at F gives the construction at the knee of the portal and eaves of the roof. The 275 mm cavity wall (with its inner face 13 mm clear of the portal has tile creasing beneath the gutter. The latter is framed with 40 mm timbers carried on 75 mm by 50 mm bearers at 400 mm centres and is lined with copper. The backs of the bearers have 75 mm wide expanded metal lathing nailed to the top and bent down into the brickwork.

The roof is of 75 mm reinforced wood-wool slabs spiked down to the wall plate and purlins. The slabs are brushed with cement slurry to give a suitable surface for the built-up felt roofing. The ceiling is of aluminium foil backed plasterboard skimmed with plaster.

At F the purlins are fixed to blocks and an alternative method of supporting them is given at H. M.s. hangers are used which are screwed and recessed flush with the top of the portal ; side coach bolts fasten the purlin ends to the hanger.

[1] Like A, Fig. 154 but with teeth on one side only.

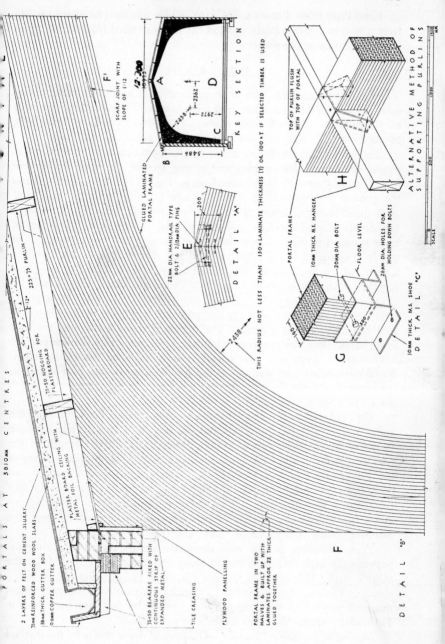

PORTALS AT 3810MM CENTRES

2 LAYERS OF FELT ON CEMENT SLURRY

75MM REINFORCED WOOD WOOL SLABS

38MM THICK GUTTER BOX

600MM COPPER GUTTER

75×50 BEARERS FIXED WITH CONTINUOUS STRIP OF EXPANDED METAL

TILE CREASING

PLYWOOD PANELLING

PLASTER BOARD CEILING WITH METAL FOIL BACKING

75×50 NOGGING FOR PLASTERBOARD

225×75 PURLIN

1'–12"

SCARF JOINT WITH SLOPE OF 1:12

F'

GLUED LAMINATED PORTAL FRAME

KEY SECTION

12·300
10·975

A

B

C

D

7438

2438

2862

2972

5486

PORTAL FRAME IN TWO HALVES & BUILT UP WITH LAMINATES APPROX 22 THICK GLUED TOGETHER

F

DETAIL "B"

THIS RADIUS NOT LESS THAN 150×LAMINATE THICKNESS (t) OR 100×t IF SELECTED TIMBER IS USED

7438

DETAIL "A"

22MM DIA. HANDRAIL TYPE BOLT & 2/15MM DIA. PINS

E

200

ALTERNATIVE METHOD OF SUPPORTING PURLINS

TOP OF PURLIN FLUSH WITH TOP OF PORTAL

PORTAL FRAME

H

1500MM

1000

500

SCALE

PORTAL FRAME

10MM THICK M.S. HANGER

20MM DIA. BOLT

FLOOR LEVEL

100

100

26MM DIA. HOLES FOR HOLDING DOWN BOLTS

G

10MM THICK M.S. SHOE

DETAIL "C"

SCALE

KING POST ROOF TRUSS (A, Fig. 116).—This was used for spans varying from 5 to 9 m ; the distance apart of the trusses varied from 2·4 to 3·7 m centres. It is now obsolete.

SHELL ROOFS

These roofs comprise two or more layers of t. and g. boarding acting both as cladding and structure over wide spans. They can take the form of North-light or barrel vault roofing, portions of cones or hyperbolic paraboloids as in Figs. 117 and 118. Such are known as *shell roofs*.

The hyperbolic paraboloid shape results by taking a building which is square or rectangular on plan and raising two of the opposite corners (*e.g.* corners O and R^1 in the sketch at B, Fig. 117). This sketch shows the general arrangement of a 9 m square building with the roof clad with three layers of boarding covered with roofing felt. The first layer boards are stapled together and laid parallel to one side. The second layer is nailed to the first layer and placed parallel to the other side. The top layer is nailed to the other two and laid diagonally. In addition, for a width of 1·2 m all round the perimeter the layers of boarding are glued together with a waterproof adhesive. From the detail at J it will be observed that the edges of the roof are stiffened all round by a glues, laminated beam (250 mm by 150 mm nominal) in which the three layers of roof boarding are sandwiched. The details H and K show how the roof is anchored down to the corner columns ; note that across the low corners only there is a 30 mm dia. tension bar to stabilize the construction. Rainwater outlets are provided at the low corners as shown at H and K, the latter also shows the stone cap to the column (in two pieces) over which the roofing felt is dressed.

Fig. 118 shows the geometry of the roof. D is a diagrammatic plan and A a diagrammatic elevation looking at side OR. At A the rise has been increased above the normal amount (one-quarter of the length of the roof side) so as to show the geometry more clearly. The plan at D shows the 9 m square roof divided by lines. A feature of the roof shape is that these lines are straight on plan and in section despite the fact that the roof is curved in two directions. That the lines are straight is clearly shown at A ; this greatly simplifies the formwork required to erect the roof (the formwork being straight for all lines parallel to the sides).

True elevations of the roof diagonals are parabolas; the diagonal between the high corners O and R^1 being concave when viewed from the outside whilst that between the low corners 6 and R being convex when similarly viewed. On the other hand certain horizontal sections through the roof parallel to the diagonals are hyperbolas (*e.g.* see note *re* E in Fig. 118). Hence the name hyperbolic paraboloid.

Method of drawing a Parabola.—One way of doing this is by the method

KING POST ROOF

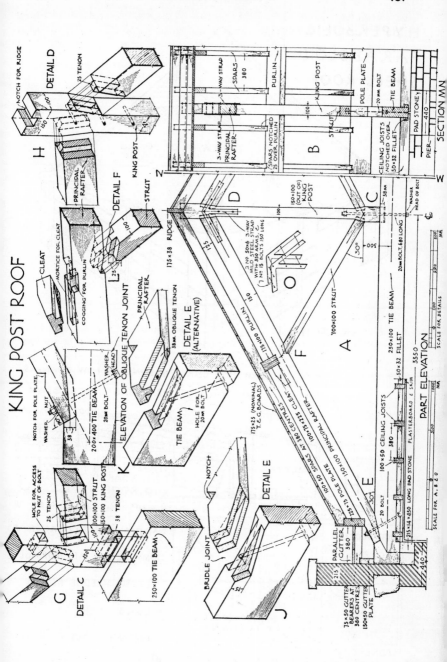

DETAIL D — NOTCH FOR RIDGE — 25 TENON — 150 100 — PRINCIPAL RAFTER — KING POST

DETAIL F — CLEAT — MORTICE FOR CLEAT — COGGING FOR PURLIN — 100 — 25 — STRUT

DETAIL H — PRINCIPAL RAFTER — KING POST — STRUT

ELEVATION OF OBLIQUE TENON JOINT — KING POST — PRINCIPAL RAFTER — 58mm OBLIQUE TENON — TIE BEAM — HOLE FOR 20mm BOLT

DETAIL E (ALTERNATIVE) — 58mm OBLIQUE TENON — 115×100 PURLIN — 115×100 PRINCIPAL RAFTER — 150 — 125

DETAIL K — NOTCH FOR POLE PLATE — WASHER — NUT — 58 — 20mm BOLT — 200×100 TIE BEAM — WASHER HEAD

DETAIL G — HOLE FOR ACCESS TO NUT OF BOLT — 25 TENON — 100 100×100 STRUT — 150×100 KING POST — 38 TENON

DETAIL C — 250×100 TIE BEAM

DETAIL J — BRIDLE JOINT — NOTCH — 51

DETAIL E — 175×25 (NOMINAL) T.& G. BOARDS — 175×25 PRINCIPAL RAFTER 100/75×125 CLEAT — 100×50 SPARS AT 380 CENTRES — 115×75 POLE PLATE 150×100 PRINCIPAL RAFTER — 30° — 20 BOLT

SECTION MN — PAD STONE — 440 — PIER — 20 MM BOLT — HEAD OF BOLT — TIE BEAM — POLE PLATE — CEILING JOISTS NOTCHED OVER 50×32 FILLET

RIDGE — 175×38 RIDGE — SPARS 380 — 3-WAY STRAP — PURLIN — KING POST — PRINCIPAL RAFTER 3-WAY STRAP — SPARS NOTCHED 25 OVER PURLIN — STRUT — 100

D — 150×100 (OUT OF) KING POST — 100 — 125 — 1 No 50×6 3-WAY MILD STEEL STRAP WITH 350 ARMS. (1 No 15 BOLTS 150 LONG — 58 MM — 58 MM — WASHER — HEAD OF BOLT — 20mm BOLT 680 LONG

O — F — A — 100×100 STRUT — 30° — 300

PART ELEVATION — 250×100 TIE BEAM — 50×32 FILLET — 100×50 CEILING JOISTS — 380 — PLASTERBOARD & SKIM — 215×14×450 LONG PAD STONE — 3550 — E

SCALE FOR A, B & O — 0 500 MM — SCALE FOR DETAILS — 0 100 500 MM

75×50 GUTTER BEARERS AT 380 CENTRES — 215 — 380 — 150×50 GUTTER PLATE — PARALLEL GUTTER — 440

HYPERBOLIC PARABOLOID SHELL ROOF

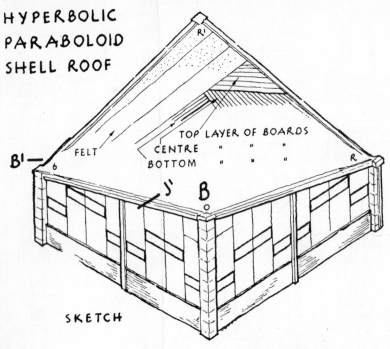

FELT

TOP LAYER OF BOARDS
CENTRE " " "
BOTTOM " " "

B¹ 6

J¹ B o

R¹

R

SKETCH

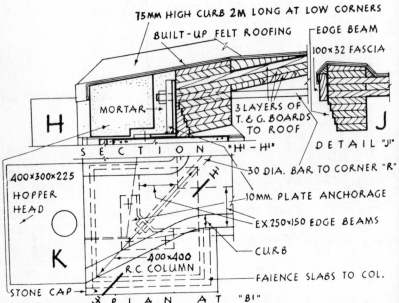

75mm HIGH CURB 2M LONG AT LOW CORNERS

BUILT-UP FELT ROOFING

EDGE BEAM

100×32 FASCIA

MORTAR

H

3 LAYERS OF T. & G. BOARDS TO ROOF

SECTION "H¹-H¹"

DETAIL "J"

J

30 DIA. BAR TO CORNER "R"

10MM. PLATE ANCHORAGE

EX. 250×150 EDGE BEAMS

CURB

FAIENCE SLABS TO COL.

400×300×225 HOPPER HEAD

K

400×400 R.C. COLUMN

STONE CAP

PLAN AT "B¹"

Fig. 117.

of tangents—see C. The height K9 is given and so is the ordinate 4K8. Produce K9 to O making 9O = K9, join 4O and O8 and divide these lines into equal parts. Draw 35, 26 and 17, these are tangent lines to the parabola which touches them midway between the intersections 7, 11, 10 and 3. Only half the curve is shown to show the construction more clearly.[1]

Method of drawing a Hyperbola.—See G. A vertical section through the apex of a cone is a triangle ; any other vertical section gives a hyperbola.

4O9 is the elevation of a cone ; 4K9 is the half plan of the cone which is cut by a vertical line ZZ. The true line of this cut on the surface of the cone is found by placing the point of the compasses at Y and describing arcs from ZZ to the base line 4Y9. Projections up from the section line ZZ will meet horizontal projections from these points at 5, 6, 7 and 8 respectively, the latter points are then on the line of half the parabola.

The following demonstrates that the true elevation of the diagonals of the roof are parabolic ; consider diagonal OR^1 on the plan at D. At F draw base line O^2R^2 parallel to OR^1 and divide it midway at S^1 ; establish point R^3 by making S^1R^3 equal to the rise 6O at A. On R^2R^3 mark off points 5^1, 4^1, 3^1, etc., to correspond with those on plan along side R^1R ; by reference to plan D it will be seen that these points coincide with Q^1, P^1 and N^1 respectively on the elevation at F. Similarly, the points L and 1, M and 2, etc., are marked on the line O^2R^3. The lines QQ^1 and 55^1 on plan appear as one line $Q5^1$ at F ; also the plan lines PP^1 and 44^1 appear as one line $P4^1$, other lines are drawn on the elevation in the same way. By scaling off half the rise ST from A, the point T^2 is marked at F as the central point on the curve of the diagonal, this will be found to meet the line $N3^1$ drawn at F. Plan points U and V on the plan diagonal are also found to give further points on the curve at U^1 and V^1 (similarly for the other intersections on the half diagonal from T to O). By joining R^2, V^1, U^1 and T^2 at F, the elevation (or sectional shape) of the half diagonal R^1T is given. It will be observed that the points T^2, U^1 and V^1 are midway between the points Y and X, X and W, and W and 5^1 respectively, therefore the lines touching the curve are tangents to it. Hence, as described above, from the tangential construction the curve is a parabola.

Now to show that certain roof sections are hyperbolic. The method of finding the plan shape of a horizontal section A^1A^1 at F is shown at E. Produce lines from D^3 and E^3 (at F) to meet the roof edges on plan D at D^1 and E^1 ; these points are the ends of the curve on plan. Two more points (F^2 and G^2) on the curve are found by projecting from the intersections (F^1 and G^1) from the plan lines at F because the section passes through these points. A curve connecting points D^1, F^2, G^2 and E^1 will give the plan shape of the cut A^1A^1 ; that this is hyperbolic is shown thus :—

Draw line H^1RH^2 at D, triangle H^1TH^2 then represents the cone (see above), leaving the section line 77 to be established. This is done by projecting a line from the known point D^1 at D to meet TH^1 at D^4 and projecting a line from D^4 to K. Placing compass point at R describe an arc from K to meet D^1D^3 at D^2, this point defines the position of section line 77. Other arcs

HYPERBOLIC PARABOLOID SHELL ROOF

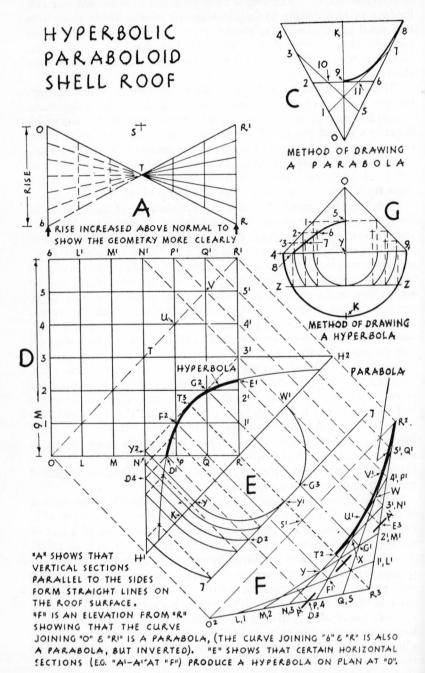

METHOD OF DRAWING A PARABOLA

METHOD OF DRAWING A HYPERBOLA

RISE INCREASED ABOVE NORMAL TO SHOW THE GEOMETRY MORE CLEARLY

"A" SHOWS THAT VERTICAL SECTIONS PARALLEL TO THE SIDES FORM STRAIGHT LINES ON THE ROOF SURFACE.

"F" IS AN ELEVATION FROM "R" SHOWING THAT THE CURVE JOINING "O" & "R" IS A PARABOLA, (THE CURVE JOINING "6" & "R" IS ALSO A PARABOLA, BUT INVERTED). "E" SHOWS THAT CERTAIN HORIZONTAL SECTIONS (E.G. "A'–A'" AT "F") PRODUCE A HYPERBOLA ON PLAN AT "D".

Fig. 118.

and projection lines give additional points on the curve showing that it is a hyperbola. For example, the projection from the plan line intersections at G^2 reaches line 77 at G^3, with centre R describe the arc G^3W^1 at E, project lines W^13^1 and 3^1 to G^2 (a point previously found on the curve). The summit of the curve at T^3 is found similarly, hence the curve found by ordinary projection is hyperbolic. Note that if a horizontal section cut were taken at F on line $N3^1$ it would pass through the plan at D from N to $T3^1$ for these points are all at the same level; such a cut would have the effect of removing one-quarter $NT3^1R$ of the roof and the plan line would be straight not hyperbolic.

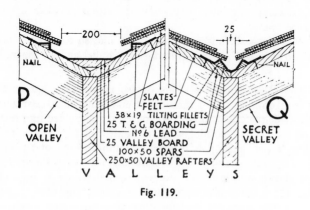

Fig. 119.

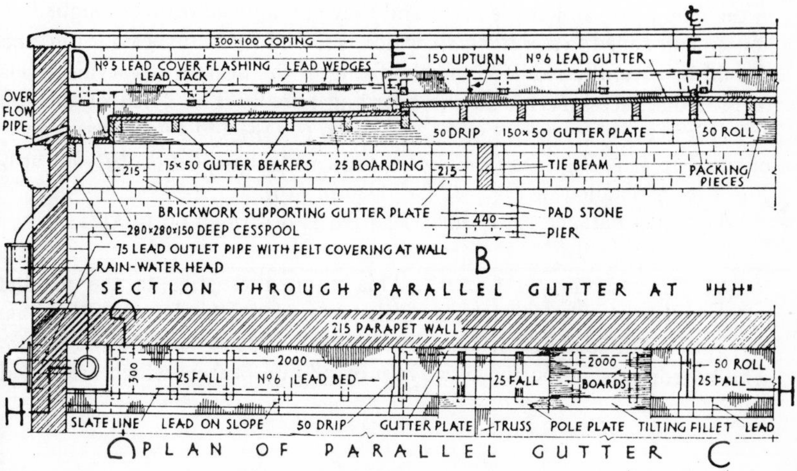

Fig. 120.

GUTTERS

The detail of the vee-gutter between two sloping surfaces (as of a pent roof) is given in Fig. 80 and described on pp. 120-122.

Two valley gutters are shown in section in Fig. 119. The ends of the spars (jack rafters) are nailed to the valley rafters, and the boarding is mitred over them. P shows an *open valley gutter* (*i.e.*, the clear width between the edges of the slates is at least 200 mm) ; a 25 mm thick *valley board* is used to block out the angle, and a tilting fillet is fixed at each side. The *secret valley gutter* at Q has tilting fillets but no valley board. If a roof is battened to receive the slates (or tiles) but not boarded, it is necessary to fix a 230 mm wide board (called a *lier board*) on each side of the intersection and for the full length of the valley to receive the lead (and tilting fillets) ; the ends of the slating battens are then cut to the edges of these boards.

In Fig. 112 the wall finishes as a parapet, and so a gutter is needed. There are two forms of parapet gutters—(*a*) parallel and (*b*) tapered.

(a) *Parallel Parapet Gutter.*—This is also known as a *box* or *trough* gutter. It is of uniform width throughout and must be at least 250 mm wide to afford adequate foot room. A long gutter is divided into sections, having a roll (p. 118) at the highest point, and drips (p. 118) at intervals not exceeding 2·5 m apart ; it is given a minimum fall of 13 mm per 100 mm (p. 113). A part plan and longitudinal section of the gutter at E, Fig. 116, is shown in Fig. 120. An enlarged detail of E, Fig. 116, is given in Fig. 121 ; this is also a section through GG, Fig. 120.

The feet of the spars are birdsmouthed to a horizontal beam, called a *pole plate* (Fig. 121), which is notched out and spiked to the principal rafter. The 25 mm boarding, laid to falls, is supported by 75 mm by 50 mm *gutter bearers* at 380 mm centres which at one end are tongued and nailed to the pole plate, and at the other end they are notched over or tongued or housed and nailed to a *gutter plate* which is spiked to the tie beam. Fig. 120 should be carefully studied, and the fixing of the bearers at varying levels to give the necessary fall to the boarding noted. The detail of the roll at F, Fig. 120, is given in Fig. 122 (see also p. 118) and Fig. 123 is a detail of the drip at E, Fig. 120 (see also p. 118). The cesspool or *drip-box* at D, Fig. 120, is detailed in Fig. 124 (see also p. 119).

A parallel parapet gutter for a lead flat is described on p. 118.

(b) *Tapered Parapet Gutter.*—This is an alternative to the parallel gutter and is so called because of its shape on plan. As shown in Fig. 125, this gutter is divided into sections by a roll and drips as described on p. 118. The boarding is supported by 50 mm by 38 mm bearers which are nailed

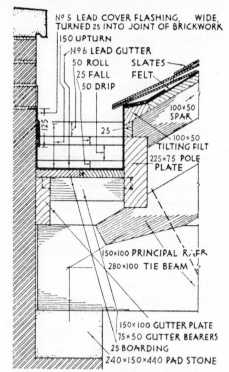

N° 5 LEAD COVER FLASHING, WIDE,
TURNED 25 INTO JOINT OF BRICKWORK

150 UPTURN

N°6 LEAD GUTTER

50 ROLL   SLATES

25 FALL   FELT

50 DRIP

125

25

100×50 SPAR

100×50 TILTING FILT

225×75 POLE PLATE

150×100 PRINCIPAL RAFTER

280×100 TIE BEAM

150×100 GUTTER PLATE
75×50 GUTTER BEARERS
25 BOARDING
240×150×440 PAD STONE

SECTION THROUGH PARALLEL GUTTER
AT "GG"

Fig. 121.

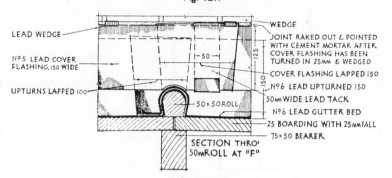

LEAD WEDGE

N° 5 LEAD COVER
FLASHING, 150 WIDE

UPTURNS LAPPED 100

50

125

150

50×50 ROLL

WEDGE

JOINT RAKED OUT & POINTED
WITH CEMENT MORTAR AFTER
COVER FLASHING HAS BEEN
TURNED IN 25MM & WEDGED

COVER FLASHING LAPPED 150

N°6 LEAD UPTURNED 150

50mm WIDE LEAD TACK

N°6 LEAD GUTTER BED

25 BOARDING WITH 25MM FALL

75×50 BEARER

SECTION THRO'
50mm ROLL AT "F"

Fig. 122.

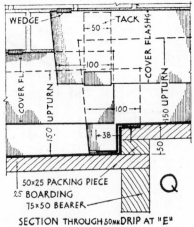

SECTION THROUGH 50ᴹᴹ DRIP AT "E"

Fig. 123.

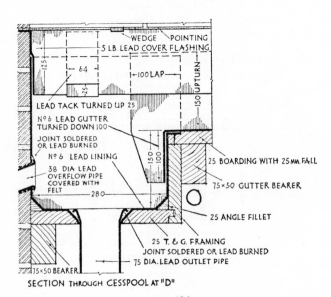

SECTION THROUGH CESSPOOL AT "D"

Fig. 124.

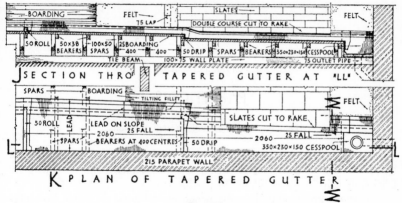

SECTION THRO' TAPERED GUTTER AT "LL"

PLAN OF TAPERED GUTTER

Fig. 125.

at varying levels to the sides of the spars and 50 mm by 38 mm uprights which are half-lapped to the bearers. The enlarged cross section in Fig. 126 shows this construction. The tilting fillet is fixed with its lower edge 75 mm above and parallel to the intersection between the gutter and roof boarding. The details of the roll, drips and cesspool are similar to those already described.

Chimney Stack Gutter.—A small gutter is required at the back of a chimney stack, such as that in Fig. 96. As shown at E, the angle at the intersection is blocked out by a triangular piece of wood which is shaped and given a slight fall in both directions from the centre. The small tilting fillet (often omitted) should also be tapered.

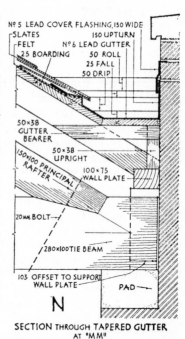

SECTION THROUGH TAPERED GUTTER
AT "MM"

Fig. 126.

NOTE.—The leadwork and slating associated with these gutter details have not been fully described as they are the chief concern of the craftsmen responsible for the work. They have been included here in order that the woodworking student may have some knowledge of the work necessary to make roofs watertight.

CHAPTER SIX

PARTITIONS

Details of stoothed and framed partitions.

TIMBER is only one of many materials (including clay and terra-cotta blocks, concrete and plaster slabs, asbestos-cement sheets and glass bricks) used in the construction of partitions.

Partitions are essentially walls which are adopted to divide buildings into rooms, corridors and cubicles ; as such they are usually thin and of light construction. Occasionally they may be required to support floor joists, ceiling joists, purlins, etc. ; such are then of heavier construction.

There are two kinds of timber inner partitions, *i.e.* (1) stoothed and (2) framed.

1. STOOTHED PARTITIONS.—These are also called *stoothings*, or *stud*, *quarter* or *common partitions*. This type is shown in Fig. 127. It consists of vertical members known as *studs* or *quarters* ; these are secured to two horizontal lengths of timber, the upper being the *head* and the lower the *sill*. It is usually covered on one or both sides with plasterboard and plaster, and occasionally with t. and g. boarding (sometimes vee-jointed —see Fig. 104), plywood sheets (Chapter Two) and wall boards (p. 170).

The studs are usually of 100 mm by 50 mm stuff ; occasionally their size is 75 mm by 50 mm. Their distance apart depends upon the covering, *i.e.*, 350 to 400 mm for lathing and plaster, and up to 600 mm centres for board-ing, plywood sheets and wall boards. Their ends may be either stub-tenoned into the head and sill (see D, E and L), or housed (let in full width), or, as shown at J, slotted over 38 or 25 mm by 25 mm fillets nailed to the head and sill. The sill is omitted in cheap work, when the studs are then nailed direct to the floor (see U). The studs are stiffened by *nogging pieces* or *noggings* at 90 to 120 cm vertical intervals. These short pieces, 100 or 75 mm by 50 or 38 mm are generally fitted more or less horizontally and tightly between the studs, to which they are nailed (see O and G), or inclined as shown at Q ; alternatively, the noggings may consist of pairs of 50 mm by 19 mm continuous pieces let in flush with the faces of the studs (see T and H). The wall studs may be packed out from the walls as shown at A, or securely plugged to the walls.

The head and sill are of the same width as that of the studs and are preferably 75 mm thick. The former is securely nailed to the ceiling (or floor) joists, and the sill is fixed to the floor. They are shown at right angles to the floor and ceiling joists. If the partition is to be fixed parallel to the joists, it should be either placed immediately over a floor joist (or

STOOTHED PARTITION

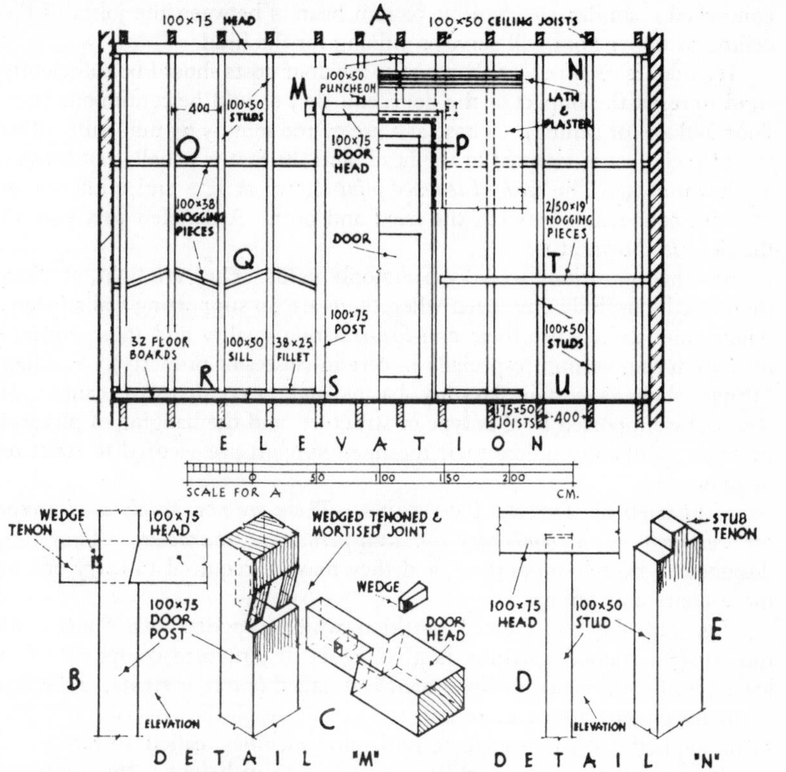

100×75 HEAD   **A**   100×50 CEILING JOISTS

400   100×50 STUDS

M   100×50 PUNCHEON

N

LATH & PLASTER

100×75 DOOR HEAD   **P**

O

100×38 NOGGING PIECES

DOOR

2/50×19 NOGGING PIECES

Q   **T**

100×75 POST

100×50 STUDS

32 FLOOR BOARDS   100×50 SILL   38×25 FILLET

R   **S**   **U**

175×50 JOISTS   400

E L E V A T I O N

SCALE FOR A   0   50   100   150   200   CM.

WEDGE TENON

100×75 HEAD

WEDGED TENONED & MORTISED JOINT

WEDGE

DOOR HEAD

100×75 DOOR POST

B

ELEVATION

C

100×75 HEAD   100×50 STUD

STUB TENON

D   ELEVATION   **E**

D E T A I L "M"   D E T A I L "N"

75×19 ARCHITRAVE
150×38 CASING
100×75 DOOR POST
3-PLY PANEL

PLASTER   LATH

STILE

F

D E T A I L "P"

100×50 STUDS
100×38 NOGGING PIECE

50×19 NOGGINGS

G   **H**

100×50 STUD
38×25 FILLET
100×50 SILL

100×75 DOOR POST

100×50 STUD

D E T A I L S "O" & "T"

J   **K**

100×75 SILL

L

A L T E R N A T E D E T A I L "R"

D E T A I L S "R" & "S"

SCALE FOR DETAILS   0   100   200   MM.

Fig. 127.

doubled joist) or, where this is not possible, on short 150 mm by 50 mm transverse bearers housed at about 90 cm centres between the pair of joists concerned ; similar 100 mm by 50 mm bearers between the joists of the ceiling or upper floor will serve as a fixing for the head.

If a door is required, as shown at A, the door posts should be sufficiently rigid to resist the impact of the door, and they should be continuous from floor to floor (or ceiling). Those at A are of 100 mm by 75 mm stuff. The top of each post is tenoned to the head, and the foot is usually *slot tenoned* as shown at K. The *wedged tenoned joint* shown at A, B and C affords an effective connection between the head and post. A detailed part plan of the door is shown at F.[1]

Stoothed partitions are very commonly used. They are light, and can therefore be usefully employed when there are no supporting walls below. Their chief defects are their non-fire-resisting quality and their capacity for harbouring vermin (especially in certain classes of property). Sanitary fittings such as certain lavatory basins and heavy fixtures, cannot, of course, be supported by this type of structure and the hanging of pictures presents a difficulty unless their means of support are secured to studs or noggings.

2. FRAMED OR TRUSSED PARTITIONS.—*These are now rarely used except for buildings of a temporary or semi-permanent character.* They are designed to be self-supporting, and they may be required to carry one or more floors and ceilings.

Fig. 128 shows a framed partition which supports two floors. As indicated, a framed partition is a triangulated structure composed of at least a head, sill, posts, inclined members called *braces* or struts, and studs (with noggings) spaced according to the nature of the covering material to be applied. An intermediate horizontal member, called an *intertie*, is usually provided when a partition exceeds 3 m in height ; this increases the rigidity of the structure, it has the effect of reducing the length and consequently the sizes of the braces and posts, and transmits a portion of the load direct to the walls. In addition to these wood members, this partition has two long wrought iron or mild steel rods which pass through the sill, intertie and head. Both ends of each rod are provided with nuts and washers. The object of these rods is to ensure close joints between the members when the nuts are tightened.

The braces, being in compression, assist in transmitting the weight to the walls. The sill, like a tie beam of a king post roof truss (see p. 157), forms a tie and is in tension. The head acts as a straining beam for the braces, forms a fixing for the posts and studs, and supports the floor joists which are cogged to it (resembling H, Fig. 66). Satisfactory bearings for the sill, head and intertie are obtained if their ends are supported on 75 to 150 mm thick hard stone or concrete pads, as shown.

The left side of the partition shown in Fig. 128 indicates a typical arrangement of the studs when the partition is to be plastered. The right side shows the partition boarded ; for this and similar treatment (see

Doors are fully described and detailed in the companion volume " Joinery."

p. 166) the studs and noggings are so spaced to provide the necessary means for fixing the covering.

There are several methods of setting out the members at the joints, but the centre line principle is usually adopted, as shown in the details in Fig. 129.

Two alternative methods of connecting the upper brace, post and head, are shown at C and D, Fig. 129. The former shows the centre line principle of setting out, the brace and post being respectively bridle jointed (Fig. 116) and morticed and tenoned. The detail at D shows an additional member, a *straining piece*, which is sometimes introduced to increase the rigidity at the heads of the posts.

The connection between the two portions of the post, the lower brace and the intertie is detailed at E. The studs are stub-tenoned into the intertie and notched over 25 mm square fillets at the brace (see G); in cheap work the ends of the studs are cut to the required bevel and nailed to the brace.

In the detail at F, the foot of the brace is bridle jointed to the sill, and a 12 mm diameter wrought iron bolt is provided to make a rigid connection. The *seating block* shown is preferred to the alternative of notching

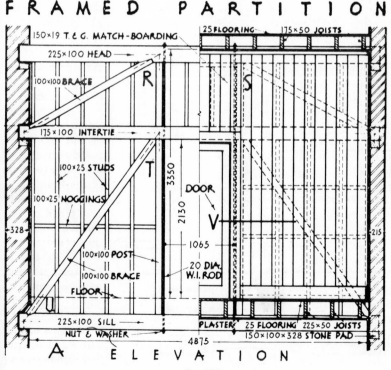

FRAMED PARTITION

150×19 T. & G. MATCH-BOARDING
25 FLOORING   175×50 JOISTS
225×100 HEAD
100×100 BRACE
R
S
175×100 INTERTIE
100×25 STUDS
T
3350
100×25 NOGGINGS
DOOR
2130
100×100 POST
V
328
215
1065
100×100 BRACE
20 DIA. W.I. ROD
FLOOR
225×100 SILL
PLASTER   25 FLOORING   225×50 JOISTS
NUT & WASHER
150×100×328 STONE PAD
4875
A   ELEVATION

Fig. 128.

DETAILS OF FRAMED PARTITION

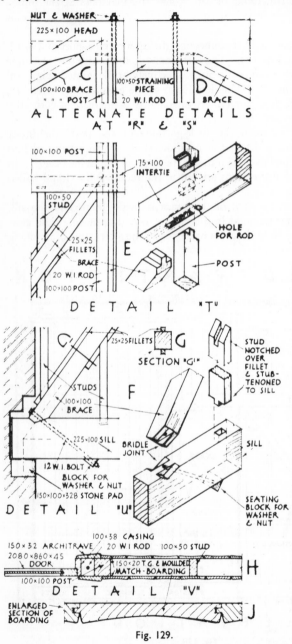

NUT & WASHER
225×100 HEAD
100×100 BRACE
" " POST
100×50 STRAINING PIECE
20 W.I. ROD
BRACE
C
D

ALTERNATE DETAILS
AT "R" & "S"

100×100 POST
175×100 INTERTIE
100×50 STUD
25×25 FILLETS
BRACE
20 W I ROD
100×100 POST
HOLE FOR ROD
POST
E

DETAIL "T"

25×25 FILLETS
G
SECTION "G'"
STUDS
100×100 BRACE
225×100 SILL
12 W.I. BOLT
BLOCK FOR WASHER & NUT
150×100×328 STONE PAD
F
STUD NOTCHED OVER FILLET & STUB-TENONED TO SILL
BRIDLE JOINT
SILL
SEATING BLOCK FOR WASHER & NUT

DETAIL "U"

100×38 CASING
150×32 ARCHITRAVE
2080×860×45 DOOR
100×100 POST
20 W I ROD
100×50 STUD
150×20 T.G. & MOULDED MATCH-BOARDING
H
ENLARGED SECTION OF BOARDING
J

DETAIL "V"

Fig. 129.

the underside of the sill (as shown at E and J, Fig. 116) for the washer, as notching would unduly weaken the sill. The section at G shows the 25 mm square fillets which are nailed to the brace and over which the ends of the studs are notched.

Just before the covering material is applied or fixed, the tension rods are given a final tightening by applying a spanner to the nuts. This closes any joints which may have opened slightly during the erection of the partition and owing to shrinkage of the timbers.

A sectional plan through a portion of the door, etc. is shown at H, and an enlarged section of the boarding is shown at J. Ordinary t. and g. vee-jointed boarding (Fig. 104) in narrow widths, wall boarding (p. 171), etc. may be preferred. Small blocks are packed between the door casing and the posts and intertie to which the casing is securely fixed.

CHAPTER SEVEN

SOUNDPROOFING

Insulating materials and their application to floors and partitions.

It is desirable, as far as possible, to prevent the transmission of sound from one part of a building to another. This applies particularly to buildings of a domestic character such as houses.

Sound may be air-borne (examples being speech and music), or it may originate by direct contact (such as noises produced when walking, hammering or banging doors, and known as impact sound). It travels considerable distances along solid walls, and is thereby transmitted from one room to another. The prevention of sound transmission through parts of a structure is known as *sound insulation*.

Many so-called *insulating materials* are available for application to walls, partitions, floors and ceilings. Because of their porous nature, they absorb sound and partially reduce the volume transmitted. These insulating materials include slag wool, wall boards, Cabot's quilt and felt.

Slag wool is a very light fibrous fireproof material produced by the blowing of steam through molten blast-furnace slag. It is obtainable either in a loose condition, or in quilt or blanket form, or in slabs. The quilted kind has slag wool sewn between brown paper and is obtainable in 90 cm wide rolls of 20 to 25 mm in thickness. The slabs have a 13 mm thick facing of plaster and a backing of slag wool which is at least 25 mm thick.

There are several proprietary names applied to *wall boards* (of wood fibre, or shavings), including *Celotex, Insulite, Lloyd Board and Tentest.* Asbestos-cement wall board is another example.

Cabot's quilt consists of cured eel-grass, stitched between strong brown paper, in 90 cm wide rolls of 10 mm (single-ply), 13 mm (double-ply) and 20 mm (triple-ply) thickness.

Hair felt is a good insulating material but is not vermin proof.

The above materials, when used normally, only offer *partial insulation.* They can be applied in several layers to completely prevent the transmission of sound but only in special cases is such application made on account of its high cost.

Good quality construction is essential, as sound is readily transmitted through cracks, gaps, badly fitting doors and windows, etc. Thick carpets and rubber or cork coverings are effective in reducing the amount of sound transmitted through floors.

The following are typical examples showing the application of sound insulation materials to wood floors and partitions.

Fig. 130 shows two sections through part of a floor. That at H indicates a method which has been employed for many years. The insulating material is slag wool, of 50 to 75 mm thickness, on plywood or rough boarding supported on fillets nailed to the joists. This treatment is very inadequate, as sound is transmitted to adjacent rooms through the thin

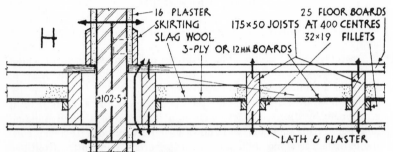

SECTION SHOWING INADEQUATE INSULATION OF FLOOR & WALL. SOUND PATHS ARE INDICATED BY THICK ARROWS

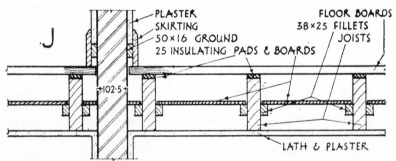

THIS FLOOR INSULATION IS MORE EFFECTIVE THAN THAT AT 'H', BUT TRANSMISSION OF SOUND THROUGH WALL NOT REDUCED

Fig. 130.

brick wall, at the gap between the wall and the wall joist, and at each joist because of the direct contact between it and the floor boards and ceiling (see thick arrows). The insulation would have been more effective if, as shown at J, narrow strips of insulating board had been lightly nailed to the joists before the floor boards had been fixed. J shows insulating boards supported on but not nailed to fillets; instead of the fillets, narrow 13 mm thick strips of the insulating material may be nailed to the joists. Alternatively, Cabot's quilt (preferably 3-ply) may be used to cover the whole floor (well lapped and nailed to the top of the joists before the floor boards are fixed) in lieu of the insulating pads and boards.

A more effective but expensive method is shown in Fig. 131. Three layers of insulating boards are used. Quilt, such as that shown at P, Fig. 133, may be used instead of the top layer of insulating boards. Impact sound (p. 171) is isolated from the wall if, as shown, the skirting is kept

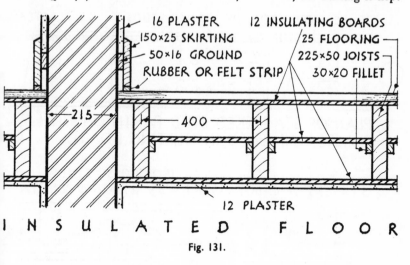

16 PLASTER   12 INSULATING BOARDS
150×25 SKIRTING   25 FLOORING
50×16 GROUND   225×50 JOISTS
RUBBER OR FELT STRIP   30×20 FILLET

215

400

12 PLASTER

INSULATED FLOOR

Fig. 131.

clear of the floor and bedded on a rubber, asbestos or felt insulating strip. Alternatively, as indicated at L, Fig. 132, the bottom edge of the skirting may be bevelled to minimize contact between it and the floor.

The section in Fig. 132 shows an excellent (but costly) method because of the separate ceiling and the insulating layers at the floor and wall; the vertical insulating boards are secured to 50 mm by 20 mm wood *grounds* plugged horizontally to the wall at intervals.

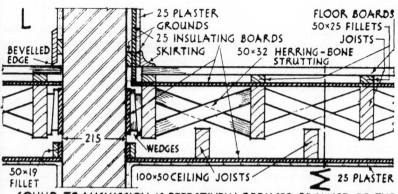

L

25 PLASTER   FLOOR BOARDS
GROUNDS   50×25 FILLETS
25 INSULATING BOARDS   JOISTS
SKIRTING   50×32 HERRING-BONE STRUTTING

BEVELLED EDGE

215

WEDGES

50×19 FILLET   100×50 CEILING JOISTS   25 PLASTER

SOUND TRANSMISSION IS EFFECTIVELY REDUCED BECAUSE OF THE INDEPENDENT CEILING & THE INSULATION OF THE FLOOR & WALL

Fig. 132.

Three plans showing alternative treatment at partitions are given in Fig. 133. Detail O shows a modified form of stoothed partition (p. 166). The discontinuity produced by the staggered studs is partially effective even if the insulating boards shown are not used and the

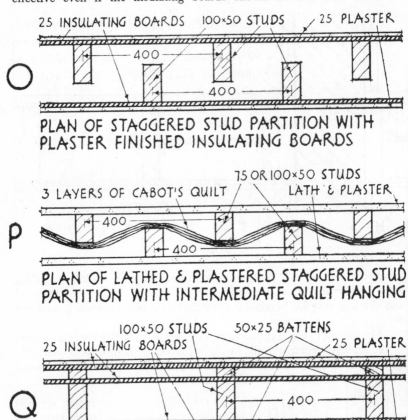

O

25 INSULATING BOARDS   100×50 STUDS   25 PLASTER

400

400

PLAN OF STAGGERED STUD PARTITION WITH PLASTER FINISHED INSULATING BOARDS

P

3 LAYERS OF CABOT'S QUILT   75 OR 100×50 STUDS   LATH & PLASTER

400

400

PLAN OF LATHED & PLASTERED STAGGERED STUD PARTITION WITH INTERMEDIATE QUILT HANGING

Q

100×50 STUDS   50×25 BATTENS

25 INSULATING BOARDS   25 PLASTER

400

PLAN OF STUD PARTITION SHOWING FOUR LAYERS OF INSULATING BOARDS AND PLASTER FINISH

ALTERNATE DETAILS

Fig. 133.

studs are simply lathed and plastered. That at P shows quilt nailed to the studs; this is a very good type. An effective but expensive sound insulated partition is shown at Q; four layers of insulating material are incorporated; floors may be also constructed on this principle.

Four typical details at the head and sill of stud partitions are shown in Fig. 134. The letters J, L, Q and X mentioned in the description in this figure refer to Figs. 130, 132, 133 and 135 respectively. The precautions taken to isolate the partitions from the floors should be noted; thus, the

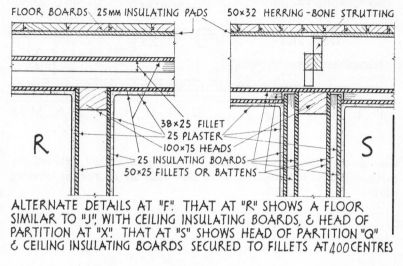

ALTERNATE DETAILS AT "F". THAT AT "R" SHOWS A FLOOR SIMILAR TO "J", WITH CEILING INSULATING BOARDS, & HEAD OF PARTITION AT "X". THAT AT "S" SHOWS HEAD OF PARTITION "Q" & CEILING INSULATING BOARDS SECURED TO FILLETS AT 400 CENTRES

ALTERNATE DETAILS AT "E". THAT AT "T" SHOWS SECTION "M" OF FLOOR "L" & SILL OF PARTITION "X". THE PARTITION AT "U" IS SHOWN PARALLEL TO THE JOISTS & INSULATED FROM THE FLOOR

Fig. 134.

insulating pad at U is wider than the wood sill at the partition, and this prevents contact between the floor boards and the sill.

Doors are vulnerable to the transmission of sound from one room to another. They should be tight-fitting and as thick as possible. A special type of insulating door is shown in Fig. 135. This is known as a *flush door*. Granulated cork (a good insulator) is packed between the frame and the plywood faces (lined with asbestos) of the door. Draught strips, rubber or felt, are fixed between the edges of the door and the casing. A strip of insulating board or a pad of Cabot's quilt is fixed and packed

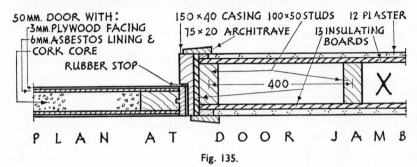

Fig. 135.

between the casing and partition. Similar pads packed between the partition (for its full height) and the side walls are effective in breaking contact.

CHAPTER EIGHT

TEMPORARY CARPENTRY

Timbering to trenches. Centering for arches up to 3 m span. Raking, flying and vertical shoring. Formwork for concrete structures.

SOME forms of timber construction are needed as temporary supports only. These include (1) timbering to trenches, (2) centering for arches, (3) shoring to walls and (4) formwork for plain or reinforced concrete structures or units.

TRENCH TIMBERING

The sides of trenches excavated in certain soils to receive wall foundations and drains must be temporarily supported. If such provision is not made, the sides may collapse, causing injury to workmen and damage to the work. Wood members are used for these temporary supports.

The timbering is not, as a rule, done by carpenters but by building labourers as the work proceeds. Nevertheless, it is necessary that the student of carpentry should have some knowledge of the nature of the work entailed, as he may be called upon to supervise it when he becomes an experienced craftsman and serves as a foreman or in a similar post of responsibility.

The sizes and arrangement of the various timbers used for such temporary supports depend upon the nature of the soil, the size of the trenches, the length of period during which the trenches are to be kept open, and the quality of the timber employed. There are many different kinds of soil, and these sometimes vary on a site. Accordingly, the spacing and arrangement of the members can only be decided on the site. The following examples should therefore be regarded as typical only.

Trenches cut in *hard ground*, such as rock and extremely hard chalk, do not require any timbering, as the sides are self-supporting.

If the ground is *firm* (as, for example, hard chalk and dense gravel) it may be necessary to provide a light support arranged as shown in Fig. 136. The vertical members, called *poling boards*, are in pairs and are strutted apart at a *minimum* distance of 1·8 m centres. This distance is required to give sufficient working space to the men

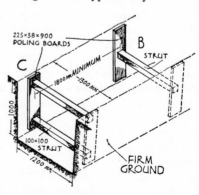

Fig. 136.

engaged in the trench either in building the wall foundations or laying a drain. One central strut, as shown at B, is usually sufficient, but occasionally two are needed (see C).

The poling boards vary from 175 to 225 mm wide by 32 to 38 mm thick by 0·6 to 1·2 m long. The struts are from 75 to 100 mm (or even larger for wide trenches) square, or they may be adjustable metal steel props of 50 to 75 mm diameter.

The struts are slightly longer than the horizontal distance between the boards and they are driven down until they are tight and more or less horizontal. The sides of the trench may be given a slight batter from the top inwards to facilitate this operation and to reduce the tendency for the members to become loose when the earth shrinks, as it does on the removal of moisture.

Hardwood (such as oak) wedges are sometimes required to tighten the struts. Thus, a wedge is driven down between an end of a strut and the adjacent poling board ; this is only necessary at one end of each strut. If struts become slack in course of time they are readily tightened by driving the wedges down as required.

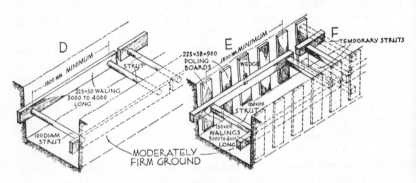

Fig. 137.

In *moderately firm ground* (such as soft chalk, loose gravel and compact clay) the sides of the trenches may be supported either as shown at D or E, Fig. 137. The former will serve if the soil is generally firm but inclined to be loose in patches ; otherwise the arrangement at E is adopted. The long horizontal members shown in both sketches are called *walings* or *waling pieces* or *planks* ; various sizes are 00 mm by 75 mm, 100 mm square, 150 mm by 100 mm, 175 mm by 75 mm, 225 mm by 50 or 75 mm by 2·5 to 4·25 m long. Those at D are thrice strutted per 4·25 m length. Wedges may be driven in between the ends of the struts, as described above. E shows poling boards held in position by strutted walings. The distance between the poling boards varies from 0·3 to 0·9 m. The timbering is fixed in easy stages, for it is not advisable in this class of soil to defer it until a length of trench is excavated equal to that of the walings, as a portion of the

unsupported excavation may collapse. Accordingly, the following procedure is adopted : A short length is dug sufficient to enable the insertion and
temporary strutting of a pair of poling boards, as shown at F (the temporary
struts being indicated by broken lines). This is repeated until sufficient
poling boards have been placed which can be spanned by the walings. A
waling is then placed along each side and strutted against the boards (see F),
after which the temporary struts are removed. Small wedges, also called
pages, are sometimes required to be driven down between a waling and
boards which have become loose ; an example of this is shown at E.

Trenches in *loose earth*, including dry sand, soft clay, ordinary loamy
soil and made-up ground (such as earth which has been tipped on to low-
lying ground and levelled off) are timbered as shown in Fig. 138, but with
the sheets placed about 25 mm apart (see below).

In a trench excavated in *loose and waterlogged ground* (such as loamy
soil and sand in which water is present) *horizontal* boards, called *sheets*, are
employed ; these
are similar in size
to poling boards,
but are from 250 to
425 cm long. These
are necessary, as it
is not possible to
dig in this class of
soil for several feet
in depth before re-
sorting to timbering.
A typical system
of timbering incor-
porating sheeting is
shown at G, Fig. 138.

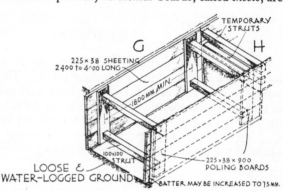

Fig. 138.

The following is the sequence of operations : The excavation is made to
a depth of about 0·3 m, a sheet is placed against each side of the excavation
and two or more struts are driven between them. The excavation is
continued for approximately 230 mm depth and a second pair of sheets is
placed tight up against the bottom edges of the first pair and strutted.
The timbering at the end of a section at this stage is shown at H. This
operation is repeated until four sets of 225 mm wide sheets have been
temporarily strutted or the required depth has been reached, when poling
boards are placed at a minimum distance of 1·8 m centres and strutted, as
shown in the sketch, and the temporary struts removed.

Measures should be taken to prevent the escape of water (and the sand
or loose soil which accompanies it) between the boards, otherwise the
members will become loose and the timbering will eventually collapse.
Hence any open joints, such as those between the ends of sheets in adjacent
sections, are stemmed by packing wads of grass, canvas, etc.

Lips are nailed at the ends of large-sized heavy struts (such as are
required for wide trenches) to facilitate the fixing of the latter. These

lips are pieces of wood, 25 to 38 mm thick, which are nailed at the ends of the struts on their top edges and project about 50 mm beyond the ends. A lipped strut is fixed simply by supporting it by the lips on the walings and driving a wedge in at the side of one end to tighten it.

Timbering to trenches for foundations is gradually removed as the construction of the wall proceeds, and when they have reached a height of two or three courses above the ground level the earth is returned to the trenches on both sides of the walls and rammed solid. Timbering to drain trenches is taken out in easy stages, and the earth is returned after the drains have been tested and approved by the local authority ; the ramming of the earth should be done carefully if damage to the drains is to be avoided.

Whilst spruce is often used for rough work of this nature, in practice old putlogs (horizontal members of scaffolding), planks, floor joists and wood which is unsuitable for better work is adopted.

CENTERING

Centres are wood members or structures which are used as temporary supports for arches during their construction. They are supported on vertical *props* or *posts* of timber. *Folding wedges* are also necessary to permit of a slight vertical adjustment of the centre to the required height at which the arch is to be constructed and for the gradual lowering of the centre after the arch has been built. These wedges are placed between the heads of the props and the centre. *Centering* includes the centre, posts and wedges.

Centres are made by the carpenter to the requirements of the bricklayer or mason. Their shape and details depend upon the type, span and width of the arches to be supported. They must, of course, be sufficiently rigid to support the weight of the brickwork or masonry to be constructed on them. Provided it is sound, any relatively inexpensive timber readily available is converted to the desired shape. The following are typical examples.

Turning Pieces.—The simplest form of centre is that required for a flat arch and is called a *turning piece*. A flat arch over a window opening is shown in Fig. 139. Despite its name, such an arch has a slight curvature or *camber* on its under surface or *soffit*. This is necessary to prevent the appearance of sagging which would be produced if the soffit were horizontal. The rise in the middle varies from 1 to 3 mm per 300 mm of span ; thus, the arch shown is given a rise of 9 mm. A turning piece is therefore a solid piece of timber having its upper surface shaped to conform to the soffit of the arch. As shown, the turning piece rests at each end upon a pair of folding wedges which are supported on a prop ; the props rest either directly on the sill or, preferably, on a horizontal member called a *sleeper* ; the latter serves as a protection to the sill. The strut, besides making the centering rigid, serves as a fixing for a nail to which the bricklayer fastens a cord used to check the accurate laying of the arch bricks (called *voussoirs*).

A turning piece suitable for an arch with a 65 mm camber is shown at D Fig. 140. This also shows the folding wedges and props.

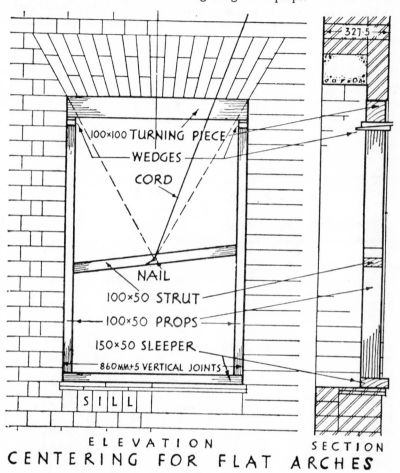

ELEVATION   SECTION

CENTERING FOR FLAT ARCHES

Fig. 139.

Centres for Segmental Arches.—Arches having wider soffits than 102 mm are " turned " upon centres which are constructed of *ribs*, *laggings* and *bearers*. A rib is a thin (usually 25 mm) piece of timber shaped to the soffit. Laggings are narrow battens (varying from 75 mm by 16 mm to 50 mm by 25 mm) nailed across the top edges of two ribs. A bearer or *bearing piece* is from 50 to 100 mm wide and 25 to 50 mm thick ; one is nailed at each end to the underside of the ribs. A centre with a similar camber to that at D, Fig. 140 is shown in the same figure at E. Both *open* and *close* lagging are shown. The former is used when the joints between the voussoirs are relatively thick (as shown in the right of the elevation in Fig. 141) ; the distance apart of the laggings varies from 20 to 25 mm for brick arches or one lagging per voussoir, and this distance is increased for masonry arches (see Fig. 143). Close lagging is adopted for what are called *gauged arches* which have very thin joints between the voussoirs (see the left half of the elevation in Fig. 141).

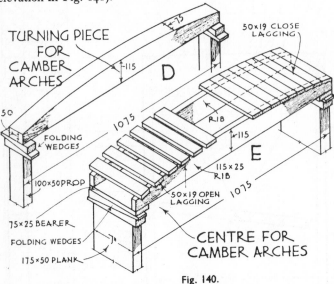

Fig. 140.

A suitable centre for a segmental arch, indicating both close and open lagging, is shown in Fig. 141.

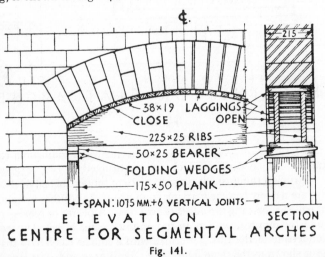

Fig. 141.

Centres for Semicircular Arches.—A suitably designed centre for a semicircular brick arch is shown in Fig. 142. To use timber which exceeds 250 mm in width is not economical and therefore it is necessary to construct the ribs as shown, with upper and lower *ties* nailed to them. Narrow laggings are used in order that they will conform to the curve. Each support consists of two props to which is nailed a 75 mm by 50 mm bearer at the top and a similar *sleeper plate* at the bottom.

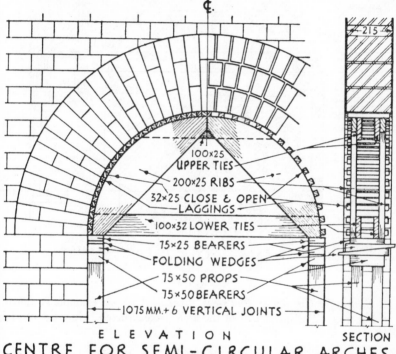

ELEVATION   SECTION

CENTRE FOR SEMI-CIRCULAR ARCHES
Fig. 142.

A centre suitable for a semicircular masonry arch is illustrated in Fig. 143. Each of the two ribs is made of two thicknesses of 230 mm by 25 mm pieces, nailed together, which overlap and have end joints that are at right angles to the curve. These are called *built-up ribs*. Each rib has double 175 mm by 25 or 32 mm ties and three 100 mm by 25 mm struts (lettered s and R), the latter being necessary to prevent deformation of the centre by the weight of the arch. The cross bracing provided by the 100 mm by 25 mm inclined brace Q and the horizontal brace T increases the rigidity. Usually two laggings per stone voussoir are provided, although these are not required if *setting wedges* (see left of the elevation) are used by the mason. Close lagging would be adopted if the arch were to be of brick. The folding wedges over the props are inserted between two stout bearers, and the props may be braced by an inclined member indicated in the section by broken lines. The *trammel rod*, required by the mason, is screwed just sufficiently tightly at the centre of the semicircle to permit it to rotate, and a *centre block* is nailed to the ties at this point.   Segmental centres up to 1·8 m span are of similar construction.

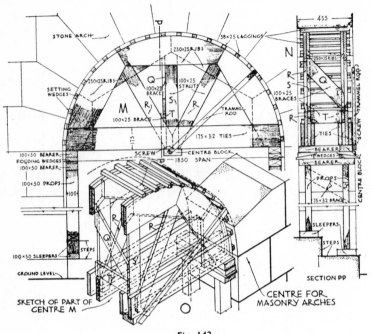

Fig. 143.

Centres for Circular or Bull's-eye Arches.—A typical example is shown in Fig. 144. The lower half of the arch is constructed before the centre

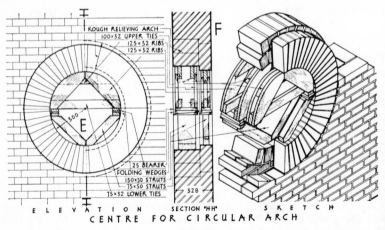

Fig. 144.

is placed in position. The centre consists of two portions, one for the external arch and the other for the internal rough arch, and it rests upon wedges supported by struts. Laggings are not required for the external arch, as it is only 102·5 mm wide ; nor are they needed for the inner arch when the span is so small. The inner portion of the centre has been omitted in the elevation.

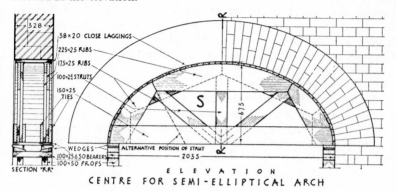

E L E V A T I O N
CENTRE FOR SEMI-ELLIPTICAL ARCH

Fig. 145.

Centres for Semi-elliptical Arches.—This type is illustrated in Fig. 145, and its construction is similar to that shown in Fig. 143. Close lagging is shown as the voussoir joints are thin ; otherwise open lagging would be adopted. As an alternative to those shown, the struts may be fixed as normals (right angles) to the inner curve (called the *intrados*) of the arch, as indicated by the relatively thick broken line.

SHORING

A *shore* is a timber member used temporarily to prop a wall which is either (*a*) defective and likely to collapse, or (*b*) liable to become so when alterations are being made to adjacent property (such as the removal of the latter subsequent to rebuilding), or (*c*) being altered by the removal of its lower portion either for reconstruction or to receive a shop front. *Shoring* is the supporting or propping of a structure with shores.

There are three types of shoring, *i.e.*, (1) raking, (2) horizontal or flying and (3) vertical, dead or needle.

1. *Raking Shores.*—These are inclined struts chiefly used to support a wall which shows signs of failure such as cracks or a bulge. Such defects may be due to thrusts from one or more upper floors or the roof (probably on account of the rafter feet being inadequately tied—see p. 122), or because of unequal settlement of its foundation.

In its simplest form this shore consists of one strut only, together with a suitable support at its foot and fixing at its head. Thus, referring to A, Fig. 146, there would be an inclined member or raking shore, supported at the ground level on a piece of wood called a *sole plate*, and secured at the top by a wood *needle* inserted in the wall. In addition, and to prevent the shore sagging, a short strut would be provided extending from the middle of the shore to the wall or wall piece (see below).

The inclination of the shore depends upon site conditions. If the wall abuts on a street, the distance between the foot and the wall may be restricted in order that traffic will not be interrupted. Preferably, the angle between the shore and the ground should be about 60° and should not exceed 75°. As far as possible, and as indicated at A, the centre line of the shore should intersect the centre of the bottom of the wall plate. The angle between the shore and the sole plate must be *slightly less* than 90° to ensure a tight fit when the foot of the shore is levered in position ; this angle is shown at A to be 87°.

Details at the head of the shore are shown at B and C, Fig. 146.

The wood *wall piece* provides a suitable abutment for the shore and fixing for the lower end of the strut or brace. It is from 150 to 225 mm wide (depending upon the size of the shore) and is usually 50 mm thick. It is holed for the needle. The wall piece is secured to the wall by means of metal (wrought iron) *wall hooks* (see B, C and J) which are driven into the joints of the brickwork (or stonework), one pair being provided near the top and bottom and at approximately 2·75 m intervals.

The *needle* (known also as a *tossle* or *joggle*) is shaped out of 100 mm square stuff and is from 300 to 350 mm long. Approximately at the intersection between the centre line of the shore and the face of the wall a brick header is removed, leaving a hole which is about 230 mm by 125 mm by 90 mm. The wall piece is then fixed and the needle inserted. The needle is strengthened to resist the upward thrust from the shore by the provision of a wood *cleat* (see B and C) which is nailed to the wall piece. Occasionally the cleat is bevel housed into the wall piece, as indicated by the broken line L at B.

The head of the shore is notched out to fit the underside of the needle (see B and C). This facilitates erection and prevents the shore from being blown down in the event of it becoming loose.

The *sole plate* or *footing block* is usually 75 to 100 mm thick and of the same width as the shore. If this bearing is inadequate (as on a yielding soil) the area is increased by using a timber *platform* of planks (say, six 225 mm by 75 mm by 900 mm long pieces) laid transversely and upon which the sole plate rests. As already stated, the sole plate must be inclined at about 87° to the shore.

After holing the wall, fixing the holed wall piece, inserting and cleating the needle, and placing the sole plate on the ground, excavated to receive it, the shore is erected with its head clasping the needle and its foot (cut to the bevel) *levered* forward by the application of a crowbar. A heavy maul must never be used for this purpose, as such may cause the building to collapse. Levering is facilitated if, as indicated at H, Fig. 146, a groove is formed in the shore foot. When the shore fits tightly against the sole plate, a wrought iron dog (K, Fig. 146) is driven into the edge of the plate and side of the shore, one at each side. Finally a shaped cleat abutting

RAKING SHORE

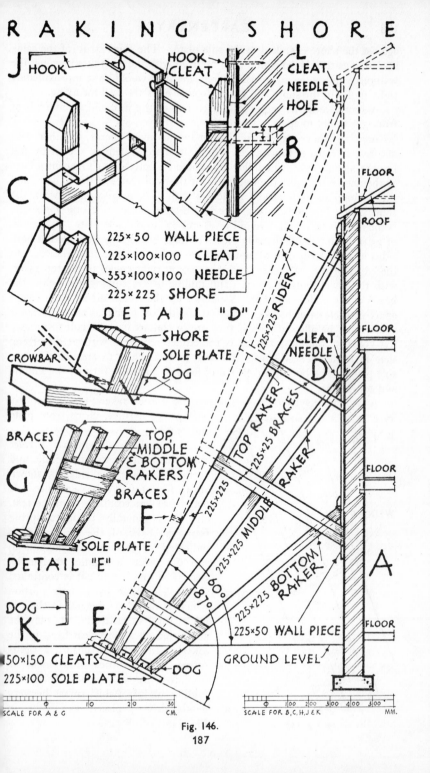

J HOOK

HOOK CLEAT

HOOK CLEAT

L CLEAT
NEEDLE
HOLE

C

B

225×50 WALL PIECE
225×100×100 CLEAT
355×100×100 NEEDLE
225×225 SHORE

DETAIL "D"

SHORE
SOLE PLATE
DOG

CROWBAR

H

225×225 RIDER

CLEAT
NEEDLE

D

FLOOR

ROOF

FLOOR

BRACES

TOP
MIDDLE
& BOTTOM
RAKERS

BRACES

G

TOP RAKER

225×25 BRACES

RAKER

FLOOR

225×225

225×225 MIDDLE

F

SOLE PLATE

DETAIL "E"

DOG

K

60°

87°

225×225 BOTTOM RAKER

225×50 WALL PIECE

A

FLOOR

E

150×150 CLEATS
225×100 SOLE PLATE

DOG

GROUND LEVEL

SCALE FOR A & G

CM.

SCALE FOR B,C,H,J&K

MM.

10 20 30

100 200 300 400 500

Fig. 146.

187

against the shore is spiked to the sole plate. The short strut (of the same scantling as the shore) would then be fixed with the ends fitted against the wall piece and underside of the shore ; the strut would be at approximately right angles to the shore and either dogs or nails are used for fixing.

A shore would be required at each end of the wall and at intervals of from 3·6 to 4·5 m apart.

A *raking* shore system consists of two or more shores, together with needles, sole plate, braces, etc. A three-shore system is shown by full lines at A, Fig. 146. The following should be noted : (1) The centre lines of the shores intersect the underside of the wall plates at the middle, and the position of each needle coincides with the intersection between the shore centre line and the wall face ; (2) the top shore (called the *top raker*) is inclined at 60° to the ground level and 87° to the sole plate ; (3) *braces*, of 25 or 32 mm stuff (usually pieces of floor boarding), are nailed at both sides of each strut (see G) and (with exception of those near the foot) to the edges of the wall piece, and these are at right angles to the top raker, with the lower ends of the top and middle braces coming just below the heads of the middle and bottom rakers ; (4) the feet of the shores are kept apart with cleats or blocks. The feet should not touch, as ample leverage space must be allowed, and if they are in contact it is difficult to remove the lower or middle raker (as is sometimes required) without disturbing that adjacent. These blocks and cleat (see sketch at G) are nailed to the sole plate. The bottom raker is first fixed, followed in turn by the middle and top rakers.

A two-shore system consists of top and bottom rakers, together with a wall piece, needles, cleats, sole plate and braces.

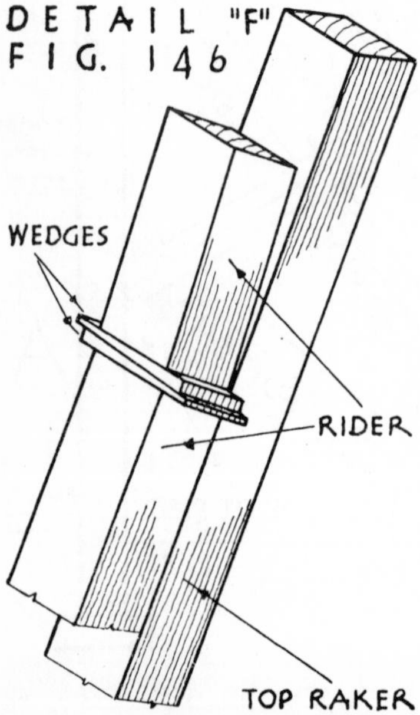

DETAIL "F"
FIG. 146

WEDGES

RIDER

TOP RAKER

Fig. 147.

A four-shore system has a *rider shore*, in addition to top, middle and bottom rakers. The rider is shown at A, Fig. 146, by broken lines. It may be of less scantling than the rest, and it is generally made to spring from the back of the top raker, as indicated ; this is more economical, and the rider is handier to handle, than if in one length ; it is continued to the sole plate by a lower portion which is nailed or dogged to the top raker (or, alternatively, it may be supported on a long cleat fixed to the raker). In order to ensure a tight-fitting rider, vertical movement is provided by a pair of hardwood folding wedges (which must be very gently driven) placed as shown in Fig. 147.

As already stated, raking shore systems are placed at from 3·6 to 4·5 m apart, with one at each end of the building. The following table gives the approximate sizes and number of raking shores per system:—

TABLE VII

Height of Wall (metres)	Number of Shores	Size of Shores (mm)
4 to 9	2	150 by 150
9 to 12	3	200 by 200
12 to 15	4	225 by 225
15 plus	4	300 by 225

Sound timber, preferably of pitch pine or Douglas fir (which are obtainable in requisite sizes and lengths), must be used for this work.

2. *Horizontal or Flying Shores.*—These are commonly applied as temporary supports to either (*a*) two gable walls adjacent to a building which is to be removed and re-built, or (*b*) a dilapidated wall abutting on a relatively narrow street and opposite which a building is available as an abutment. In these cases raking shores are unsuitable, as such would either impede building operations or interfere with traffic. Besides not requiring any ground support, flying shores are more efficient (because, as a rule, their thrust is immediately opposite to the disturbing force) and economical than raking shores.

A simple example of flying shoring is shown at A, Fig. 148. It spans a narrow street (7 m wide) and it is assumed that the upper portion of the building on the right shows signs of failure. The flying shore, which is in line with the top floor, is supported at each end on a needle strengthened by a cleat. A pair of folding wedges (of hardwood) are driven in between one of the ends of the shore and the wall piece ; this must be done carefully to avoid excessive vibration. Details showing a support and wedges are given at J and K, Fig. 148. This temporary member is strengthened, and the bearing surface increased, by the provision of two inclined (about 45°) members, called *struts*, above and below the shore ; the two top struts are similar to raking shores. The upper ends of the top struts abut against cleated needles (see F), and a horizontal member, called a *straining sill*, is nailed to the shore to restrain the feet of these struts ; each strut is tightened by inserting a pair of wedges between its foot and the sill, after which the strut is secured by a dog. These details are shown at G. The bottom ends of the lower struts abut on cleated needles, and their top ends are pressed against those of a *straining head* (nailed to the underside of the shore) when folding wedges are inserted and *carefully* driven home at the lower ends (see detail at H) ; dogs, similar to that shown at G, provide a top fixing.

The usual method of constructing these shores is as follows : Needle holes are formed in the wall, the holed wall pieces are fixed, the shore needles are inserted and cleated, the shore is raised and placed on these

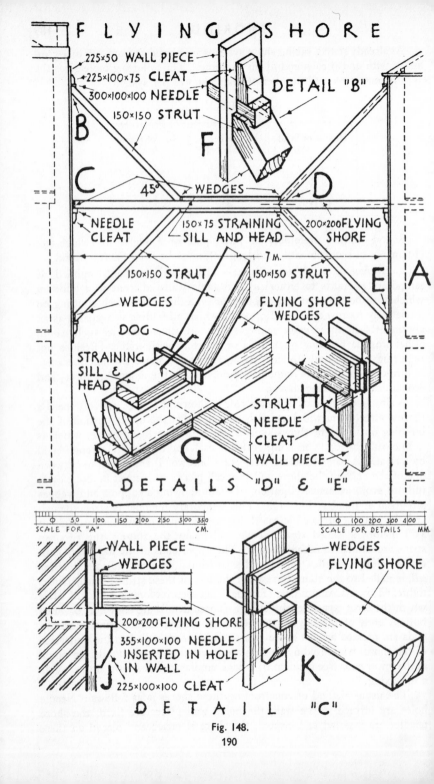

FLYING SHORE

225×50 WALL PIECE
225×100×75 CLEAT
300×100×100 NEEDLE
150×150 STRUT

DETAIL "B"

B

C   45°   WEDGES   D

NEEDLE CLEAT

150×75 STRAINING SILL AND HEAD

200×200 FLYING SHORE

7 M.

150×150 STRUT   150×150 STRUT

WEDGES

DOG

FLYING SHORE WEDGES

E   A

STRAINING SILL & HEAD

STRUT
NEEDLE
CLEAT
WALL PIECE

H

G

DETAILS "D" & "E"

SCALE FOR "A"   0 50 100 150 200 250 300 350
CM.

SCALE FOR DETAILS   0 100 200 300 400
MM.

WALL PIECE
WEDGES

WEDGES
FLYING SHORE

200×200 FLYING SHORE

355×100×100 NEEDLE INSERTED IN HOLE IN WALL

J   225×100×100 CLEAT   K

DETAIL "C"

Fig. 148.

supports and end-wedged, the lower struts are then fixed, followed by the upper.

Flying shore systems, like those of the raking class, are placed near the ends of the opposite walls and at from 3·6 to 4·5 m intervals. Pitch pine, Douglas fir (cheaper graded) and redwood are used for this class of work. Sizes of members for spans up to 12 m (which is rarely exceeded) are shown in the following table :—

TABLE VIII

Span (metres)	Size of Flying Shore (mm)	Size of Struts (mm)
Up to 4	150 by 100	100 by 100
Up to 12	150 by 150 to 225 by 225	150 by 100 to 225 by 100

3. *Vertical, Dead or Needle Shoring.*—This class of shoring is called for when (a) the lower portion of a building has become defective, probably on account of unequal settlement affecting the foundations, and the necessity arises of supporting the upper portion of the building until the foundations and defective walling have been rebuilt (known as *underpinning*), or when (b) an alteration is to be made to a sound building, such as the conversion of a house into a shop, or the replacement of an existing shop front by a modern one.

The work entailed is illustrated in Fig. 149, and the operations are carried out in the following sequence :—

(1) The windows are strutted, as shown, by 100 mm by 50 or 75 mm timbers. This is a safeguard against the windows becoming deformed if unequal settlement occurs.

(2) The roof and floors are supported by struts down to the ground (or basement) floor to remove as much weight as possible from the wall. The struts are in sections from floor to floor (or roof), one immediately over the other, each supported on a continuous sill with a continuous member at the head (see A, and F at B) ; folding wedges, carefully driven between the foot of each strut and the sill, render the struts rigid. These struts are placed at from 1·2 to 1·8 m apart.

(3) Horizontal members, called *needles*, are inserted through holes made in the wall at from 1·8 to 2·1m intervals and just above the first floor or girder (the latter being necessary if a shop front is required) ; sound brickwork or masonry will be self-supporting over this span ; in the example illustrated in Fig. 149, these needles have been placed centrally under the piers between the upper windows. Needles are generally of 300 mm square baulks, or are short lengths of mild steel beams. Each needle is supported at each end by a vertical timber, called a *dead shore*, resting on continuous sills ; the outer sill is often of baulk stuff and is known as a *sleeper*. Dead shores are usually 300 mm square baulks. The top of each shore is dogged to the needle (both sides), and the outer angle is braced.

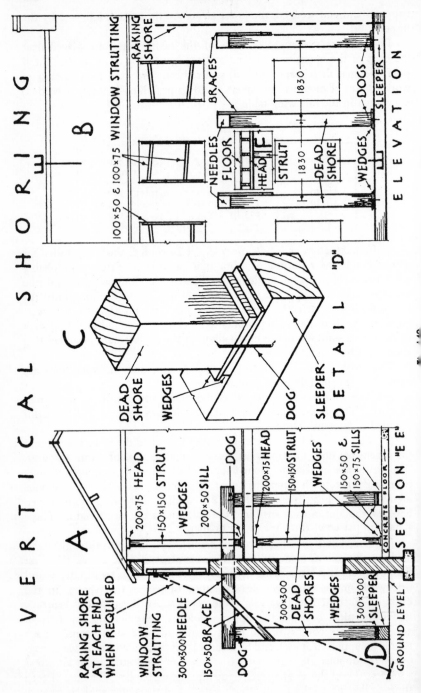

VERTICAL SHORING

A

RAKING SHORE AT EACH END WHEN REQUIRED
WINDOW STRUTTING
300×300 NEEDLE
150×50 BRACE
200×75 HEAD
150×150 STRUT
WEDGES
200×50 SILL
DOG
200×75 HEAD
150×150 STRUT
150×50 & 150×75 SILLS
WEDGES
CONCRETE FLOOR
DOG
300×300 DEAD SHORES
WEDGES
300×300 SLEEPER
GROUND LEVEL

SECTION "E"

C

DEAD SHORE
WEDGES
DOG
SLEEPER

DETAIL "D"

B

100×50 & 100×75 WINDOW STRUTTING
RAKING SHORE
BRACES
NEEDLES
FLOOR
HEAD
STRUT
1830
DEAD SHORE
1830
DOGS
WEDGES
SLEEPER

ELEVATION

Hardwood folding wedges are carefully driven at the feet of the shores to tighten the needles against the underside of the old work, after which dogs are driven in as shown at B and C. The weight of the upper portion of the wall is thus transmitted to the ground floor and ground.

(4) The wall below the needles is now removed. In the case of underpinning, the wall is re-built tight up to the old. If a shop front is required, piers are constructed at the ends of the opening and a steel girder is supported on them ; the brickwork between the girder and the old work is built, and the floor joists of the first floor are connected to the girder.

(5) A few days later, the shoring is removed in this order : (*a*) needles, (*b*) window strutting and (*c*) floor strutting. This should be done in easy stages, the wedges below the dead shores being *eased* (their points—thin edges—being knocked in slightly) and left for a day or so before being withdrawn and the shores removed.

As a further precaution against accident, raking shores, in addition to dead shores, are employed on tall or defective buildings. These are placed close to the dead shores (or near to the ends of the building only) ; they are the first to be erected and the last to be removed. One is indicated by thick broken lines at A and B, Fig. 149.

FORMWORK

This is also known as *shuttering* or *centering*, the latter term being usually applied to circular work such as arches, domical roofs, etc. Formwork is chiefly of timber construction and is therefore carried out by carpenters. Formwork is temporary work associated with the construction of concrete units or structures. In its simplest form, as required for concrete lintels, window sills, fence posts, etc., it is a wood box, called a *mould* or *casing*, shaped to the required unit and into which the concrete is cast. For reinforced concrete structures the formwork may be on a large scale to receive and support the concrete until it has sufficiently set ; thus, the formwork for a reinforced concrete floor consists of a wood platform, with wood moulds for the beams, etc. in addition to supports which transmit the loads to the floor below.

Formwork must be of adequate strength to support the concrete, and sufficiently rigid to permit of the consolidation, by ramming, of the concrete when being placed into position. It should be designed to permit of its easy and gradual removal (an operation called *striking*) either completely or partially (such as the dismantling of the side boarding before the soffit boarding of beams—see p. 196). It is also essential that the joints between adjacent sections of formwork shall be tight to prevent the escape of the liquid concrete, and therefore provision should be made to cramp or tighten the members.

The following description is that of typical details of formwork required for a simple reinforced concrete floor.

7

FORMWORK FOR COLUMNS.—Pillars or columns are often required to support the beams of a floor. The formwork consists of square edged boards or *sheeting*, placed vertically and arranged to conform to the shape of the concrete columns; therefore, the cross section may be square, rectangular, circular, octagonal, etc.

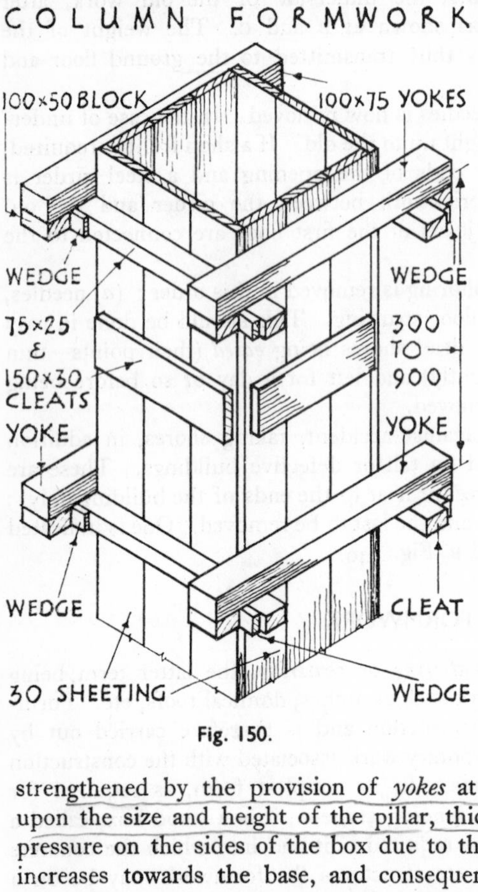

A sketch showing a common arrangement of timbering for a square reinforced concrete column is shown in Fig. 150. This is also called a *column box* or *pillar box*. The sheeting is 25, 32 or 38 mm thick Each side is formed into a panel *before* erection by nailing cross members, called *cleats,* to the boards at from 300 to 900 mm centres; this expedites assembly and removal of the formwork. The sides are maintained in position and the formwork

COLUMN FORMWORK

100×50 BLOCK · 100×75 YOKES

WEDGE · WEDGE

75×25 & 150×30 CLEATS

300 TO 900

YOKE · YOKE

WEDGE · CLEAT

30 SHEETING · WEDGE

Fig. 150.

strengthened by the provision of *yokes* at varying intervals depending upon the size and height of the pillar, thickness of sheeting, etc.; the pressure on the sides of the box due to the weight of the wet concrete increases towards the base, and consequently the distance between the yokes increases from, say, 300 mm centres at the base to 600 mm and not more than 900 mm centres at the top. That in Fig. 150 is one of several forms of yoke and consists of 100 mm by 75 mm stuff with blocks nailed at the ends. A pair of yokes grip two long cleats, and when wedges, as indicated, are driven in between the cleats and blocks, the cleats and yokes are brought tightly against the sheeting to ensure close edge joints between the sheeting panels.

The formwork is removed by easing and removing the wedges. This may take place from one to four days after the concrete (when made of ordinary Portland cement) has been placed in position.

FORMWORK FOR FLOORS.—An outline section through a typical reinforced concrete floor is shown at E, Fig. 151. The floor is divided by beams into panels or *slabs*.

Typical details of formwork, suitable for beams of relatively short span and a thin slabbed floor, are shown at A and B, Fig. 151. The formwork for the beam is a box consisting of two boarded sides, called side *shutters*, and a boarded *soffit* (bottom). Like a column box, the soffit and each side shutter are made separately into panels by nailing cleats to the boards. The thickness of the side boarding varies from 25 to 38 mm and that of the soffit boarding from 45 mm to 50 mm. This timbering is supported by *props* at, say 1·2 m centres, although this distance varies with the size and height of the props, quality of the timber, and the load which each supports. A horizontal member, called a *head-tree* or *cross-head* or *transome*, is nailed to the head of the post, and two braces, one on each side, are nailed to the prop and head-tree, as shown. The props are supported on sole pieces to distribute the load, and, as in dead shoring (p. 193), folding wedges are provided to adjust and ease the formwork.

The formwork for the floor slab is similar to the construction of an

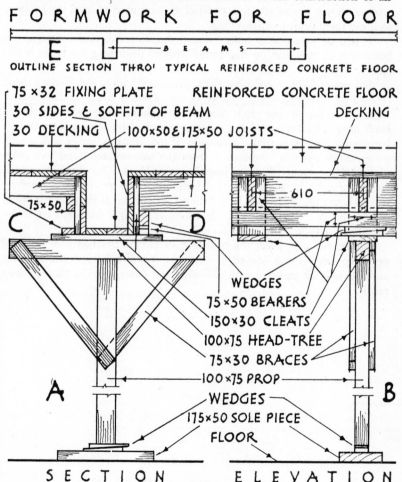

FORMWORK FOR FLOOR

E   B E A M S

OUTLINE SECTION THRO' TYPICAL REINFORCED CONCRETE FLOOR

75 × 32 FIXING PLATE   REINFORCED CONCRETE FLOOR
30 SIDES & SOFFIT OF BEAM   DECKING
30 DECKING   100×50 & 175×50 JOISTS

75 × 50

C   D

610

WEDGES
75 × 50 BEARERS
150×30 CLEATS
100×75 HEAD-TREE
75×30 BRACES
100 ×75 PROP

A   B
WEDGES
175×50 SOLE PIECE
FLOOR

SECTION   ELEVATION

7A   Fig. 151.

ordinary boarded and joisted floor in that it consists of boarding or *decking*, supported on joists or bearers. The thickness of the decking varies from 32 to 38 mm, and the sizes of the joists are from 100 to 225 mm by 50 mm and 225 mm by 75 or 100 mm. At C, 100 mm by 50 mm joists are shown supported on 75 mm by 50 mm continuous bearers nailed to the side shutter cleats. This is suitable for so-called " light floors ". The alternative detail at D, typical of formwork for a heavier floor, shows larger joists supported on bearers resting upon folding wedges placed on the long soffit cleats immediately over the head-trees (see also B). The joists or bearers span from beam to beam, and when required, have intermediate supports in the form of continuous heads propped at intervals.

The 75 mm by 32 mm *fixing plate* shown at C is nailed to the beam soffit cleats and serves to keep the beam side shutter in position. The folding wedges shown in the alternative detail at D are used to level the joists or to lower them prior to dismantling.

It will be seen that at both C and D the floor slab timbering and the beam side shutters can be dismantled without disturbing the beam soffits (see p. 193).

Softwoods, such as red pine, sitka spruce and whitewood, rough graded (known as " third or fourth quality "), are generally used for this class of work. Douglas fir and pitch pine are sometimes employed when heavy loads have to be supported and joists are required for long spans. Boarding should only be partially seasoned ; if too dry, the timber will absorb an excessive amount of moisture from the wet concrete, and will swell and buckle. Generally, the boarding should also be " thicknessed ", *i.e.* be brought to a uniform thickness, otherwise an irregular surface, due to board markings, will be shown on the surface of the concrete when the formwork is removed. The boarding is wrought (dressed) when the concrete is required to have a smooth surface.

Before being re-used, the boarding should be brushed over with oil or whitewash (or soft soaped) ; otherwise it will adhere to the concrete and will only be removed with difficulty, and damage to the concrete may result.

CHAPTER NINE

NAILS, SCREWS AND BOLTS

Wire and wrought nails, brads, panel pins, needle points, screws, corrugated saw edge fasteners, bolts, timber connectors and dogs.

THE following steel or wrought iron fastenings [1] are used in carpentry and joinery. With exception of the bolt (see Fig. 153) they are illustrated in Fig. 152.

WIRE NAILS.—These are of oval and circular section.

Oval Wire Nails (see A) are used for general purposes. They are tough and are not liable to split the timber when driven in. The slight shallow grooves or serrations in the stem increase the " holding power " or ability to grip the fibres of the wood into which they are driven. They are obtainable in sizes varying from 25 to 150 mm and are sold by weight. They are also known as *American nails.*

Circular Nails (see B), also called *French nails*, are used chiefly for temporary or unimportant work, and in the making of boxes, packing cases, etc. They are not much used by the joiner on account of the unsightly circular head.

Cut Clasp Nails (see C).—These have been practically superseded by the oval wire nail for general purposes.

Wrought Nails (see D).—As implied, these are of wrought iron, and are tapered in both width and thickness to form a point. They are usefully employed for nailing thin members, such as boards, to framing, as their points can be clenched, *i.e.*, after the nails have been driven home the points on penetrating the framing are bent over and forced into the timber by use of the punch (see I, Fig. 160) and hammer. The sizes vary from 25 to 100 mm.

Spikes are wire nails which exceed 150 mm in length and wrought nails which are longer than 100 mm. They are used in carpentry for securing large timber members.

Floor Brads (see E).—As implied, these were once used for nailing floor boards ; they have been replaced by the oval wire nail.

Joiners' Brads or Sprigs (see H) resemble floor brads, but are only from 6 to 50 mm in length. They are made of steel, brass and copper.

Panel Pins (see F).—These small nails, circular in section, are generally used by the joiner for fixing hardwood members (usually mouldings) on account of the small holes which are left.

[1] See p. 221 for special fastenings used in the cartridge-assisted tool.

Needle Points (see G) are steel pins used by the joiner for fixing small mouldings, veneers (p. 44), etc. After being driven in, they are snapped off flush with the surface. They are obtainable in six degrees of fineness.

It is difficult to drive small sprigs, pins, etc. into hardwood without bending unless holes have been previously bored to receive them. Driving is facilitated if the points are dipped in grease.

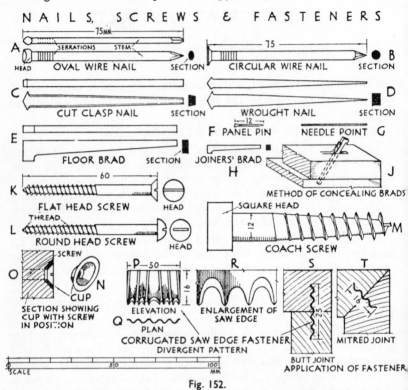

Fig. 152.

CONCEALMENT OF FASTENINGS.—In joinery especially, it is often necessary to conceal nails and other fastenings. When nails and brads are driven into softwood, their heads are forced below the surface by using a hammer on the steel punch (1, Fig. 160) and the holes are filled or "stopped" with putty (whiting ground in raw linseed oil); the woodwork is then painted.

For hardwoods which are not to be painted, the heads are punched and the holes are stopped with material which is coloured to conform to the wood. This stopping, which is melted and applied with a knife as a mastic, sets hard and is then smoothed over flush with the surface of the wood. Another method of concealing brads is shown at J, Fig. 152. A sharp chisel is used to carefully cut and lift a small portion of the wood, the brad is punched below the surface and the chip is glued down.

Pelleting is adopted for concealing screws. This consists of sinking the head below the surface after using the brace and centre bit (see 45 and 46, Fig. 159) to form a shallow hole ; a cylindrical plug or pellet of similar wood, of the same diameter as the hole, is then glued, driven in and chiselled off smooth.

SCREWS.—Of the several forms of screws, those chiefly used for fixing woodwork are (a) the *flat-headed* and (b) the *round-headed* types. These are of steel, wrought iron and brass, and are called *wood screws*, as the thread is effective in cutting into timber. Screws are fixed by means of the screw-driver (p. 215) or brace and bit (45, Fig. 159) or electric screw-driver (3, Fig. 159). Their advantages over nails are : (1) They have a greater holding power than nails, (2) they can be readily removed when required and (3) they can be fixed in positions where jarring has to be avoided.

(a) *Flat-headed or Countersunk Screws* (K, Fig. 152).—From the flat circular head (slotted to receive the screwdriver) the shaft is tapered down to a point. The *thread* proceeds in a spiral form from the pointed end to midway along the *shank*, and this threaded portion cuts into the timber as the screw is turned and inserted until the head is at least flush with the timber. It is obtainable in sizes varying from 6 to 150 mm long and from 2 to 16 mm in diameter. For hardwoods especially, it is essential to bore a hole of a smaller diameter than that of the screw by one of the several boring tools illustrated in Fig. 159 prior to inserting the screw.

Cups (see N) should be used in conjunction with screws wherever mouldings, beads, etc. are to be removed on occasion, otherwise frequent removal and re-insertion of the screws will damage the timber. They are of brass and in various sizes to suit the heads of the screws which they are to receive. A section with a cup in position is shown at O. A hole, slightly smaller than the diameter of the top of the cup, is formed by a centre bit, a little glue is placed round the hole and the cup is driven in.

(b) *Round-headed Screws* (see L) are similar to the above, except that the head is almost hemispherical. They are rarely used by the carpenter but are commonly adopted for fixing metal to wood, such as locks and similar hardware.

Coach Screws (see M).—The shanks of these resemble those of wood screws, but their heads are square or hexagonal in order that they may be turned by a spanner. Coach screws are from 30 to 200 mm long and from 5 to 12 mm in diameter. They are often used for connecting metal plates, straps and angles to timber.

CORRUGATED SAW EDGE FASTENERS (see P, Q and R).—These are corrugated pieces of steel or brass which are shaped and sharpened along one edge to give what are called *tack points*. The points are sharpened alternately on opposite sides like a saw (see R). They are from 6 to 25 mm deep and are being extensively used for making light framings, boxes and similar temporary work, repairing cracked boards, etc. Two applications are shown at S and T, Fig. 152, the former showing a butt joint and the latter a mitred joint. These fasteners are readily fixed by simply driving them in with a hammer, during which the wood members are drawn together.

BOLTS, NUTS AND WASHERS (see Fig. 153).—Bolts and nuts are commonly used in carpentry for securing members, especially of large size. Examples of structures in which they are employed include the king and built-up roof trusses (Figs. 113 and 116), built-up beams (such as the binders shown at B and C, Fig. 71) and framed partitions (Figs. 128 and 129). When bolts are used to secure wood members, washers must be introduced between the timber and the heads and nuts, to prevent the heads and nuts from being forced into the timber as the nuts are being tightened by a spanner (see K, Fig. 116, F, Fig. 129 and Fig. 153).

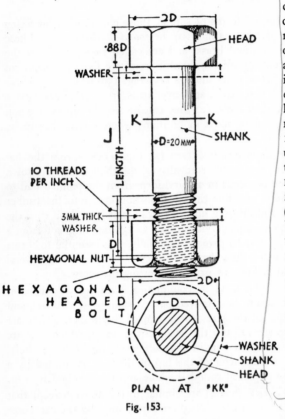

Fig. 153.

A bolt consists of a *shank* and *head*, and, as shown in Fig. 153, the proportions of the head and nut are related to the diameter of the shank. The end of the shank is in the form of a screw having a *pitch* (distance between *threads*) which varies according to the diameter of the bolt (which is that of the shank) ; thus a 6 mm diameter bolt has 20 threads to 25 mm, 25 mm bolts have 8, and a 20 mm bolt as shown has 10 threads to 25 mm. The depth of the thread varies, in the example it is approximately 2 mm. Bolts vary in size from 6 to 150 mm diameter, but rarely in carpentry is 40 mm exceeded ; the length, which is that of the shank, also varies. The thickness and diameter of a washer depend upon the size of the bolt ; that shown in Fig. 153 is 3 mm thick and the external diameter is either 40 or 45 mm. The head and nut are shown to be hexagonal on plan, and this is the type in general use ; the description of such a bolt and nut is abbreviated to " X—O—X ", meaning a hexagonal head, round shank and hexagonal nut. Square-headed bolts and nuts are sometimes adopted.

Bolts, nuts and washers are made of mild steel, wrought iron and brass, wrought iron being chiefly used in carpentry and joinery.

TIMBER CONNECTORS.—These are patent devices for strengthening timber joints, such as those between members of a built-up roof truss, described on pp. 150-153. There are several forms of connectors.

One type is called a *toothed plate*—see p. 152 and A and C, Fig. 154. It is made of steel, is from 50 to 95 mm diameter and has sharp teeth on both sides. Assuming that a simple joint between two members is required, as shown at C, a toothed ring connection is made in the following manner : With one member placed above the other, a hole is bored through both at the desired position, the top member is removed, a toothed ring is placed centrally over the hole in the bottom timber, and the top member is returned. Force is now exerted to embed the connector and bring the surfaces of the timbers together ; a special strong bolt, with nut and washers, is used for this purpose ; this bolt is placed in the hole, and after the nut has been tightened, a long handled spanner is applied to its head until the joint is closed. This special bolt is replaced by an ordinary bolt (with nut and washers). As stated on p. 152 a connector is required at each interface and thus two are needed at C.

Another type is called a *split ring*—see p. 152 and B and C, Fig. 154. This plain ring is 63 or 100 mm diameter and has a tongued and grooved break in the perimeter. A joint is made as described above, except that a special grooving tool is used to form a circular groove (half ring-depth) on the face of each adjacent timber to accommodate the ring.

DOGS have been referred to on pp. 186, 189 and 191.

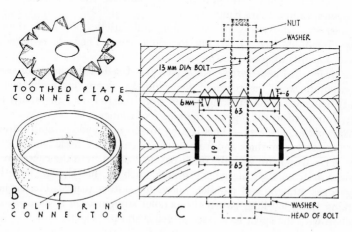

Fig. 154.

CHAPTER TEN

TOOLS

Marking and setting out tools, cutting and planing tools, boring tools, impelling tools, abrading tools, cramping and holding appliances, and miscellaneous tools and equipment.[1]   Care of tools.

THE hand tools used by the carpenter are relatively small in number. Whilst machinery has very largely displaced hand labour, particularly in shops where standardized units such as doors and windows are produced, the joiner is required to do many jobs which necessitates the use of hand tools, the number of which is fairly large, although certain of these are only needed for special work.   The use of portable power tools (operated by electricity or a small explosive charge) is now widespread ; their higher capital cost is soon recovered in greater efficiency and output of the tradesman, particularly on the building site where workshop machinery [2] is not available.   With few exceptions, the following are those which are in general use and most of them are essential parts of a kit.

CLASSIFICATION.—Hand tools may be classified into those required for : (1) Marking and setting out, (2) cutting and planing, (3) boring, (4) impelling, (5) cramping and holding, (6) abrading, (7) miscellaneous purposes, and (8) portable power tools,[3] including those in classes (2) to (4), (6) and (7), which are grouped under one classification.

1. MARKING AND SETTING OUT TOOLS (see Fig. 155).—These include rules, marking knife, straight-edge, try square, mitre square, bevel, compasses, callipers and gauges.

Rules are of boxwood.   There are several varieties, including the one-metre four-fold, (see 1), 50 cm four-fold, etc.   They are required, of course, for obtaining and setting-out sizes.

Marking Awl and Cutting Knife (see 8).—Used by the joiner for setting out accurate work (as required for windows, doors, etc.), the awl (or point) being used for pricking points from the rod, and the sharp edge being employed to cut lines on the pieces of timber.

Straight-edge (see 7).—This is a 75 to 100 mm wide carefully dressed board, 12 to 20 mm thick and from 90 to 425 cm long ; some have only one edge perfectly square with the other bevelled down from the centre to distinguish it from the true edge ; others, such as that shown, have both long edges straight and parallel ; used for testing planed, etc. surfaces, marking lines, levelling (when the parallel straight-edge is used, in conjunction with the spirit level—see sketch 7 and p. 84), etc.

---

[1] Tools used in both carpentry and joinery are included.

[2] See Chapter One, " Joinery."

[3] The electric drill is one of these and is primarily a boring tool, but by means of various attachments can be converted to a cutting, abrading and mortising tool.

Try square (see 2).—Used for setting out right angles (see p. 143) and for testing square edges during the planing up of stuff ; is obtainable with 115, 150, 190, 230 and 300 mm long blades. A larger square is also required for setting out and testing framing ; consists of a mahogany blade, which is usually 60 mm by 6 mm by 750 mm long, tenoned to a 400 mm long stock. All-metal try squares are also available, and the blades of these are graduated like rules.

A Mitre Square or Fixed Bevel has a steel blade fixed at 45° to a wood stock ; this is a useful tool for setting out 45° angles.

Bevel (see 3).—The slotted blade, which is 230, 270 and 300 mm long, can be secured at any desired angle by tightening the screw with the screwdriver ; used for setting out angles or " bevels " (see pp. 142-147) other than right angles.

Compasses (see 5).—Used for marking parallel lines to irregular surfaces, such as scribing mouldings to walls (see p. 141) and skirtings to floors, and for describing circles and setting off distances ; stocked in 150, 175, 200, 230 and 250 mm sizes.

A *trammel* (also called *beam compasses*) is used for striking large arcs or circles ; consists of two metal heads, each having a 75 to 125 mm point, which slide along a hardwood stick 30 mm by 12 mm) ; the points can be fixed as desired to the stick, and one of them may be replaced by a pencil socket.

Callipers are used for measuring diameters of curved surfaces ; *outside*

MARKING & SETTING OUT TOOLS

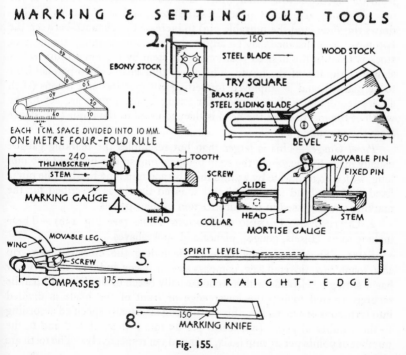

Fig. 155.

callipers, used for external dimensions, consist of a pair of hinged steel curved legs which are shaped to a fine point ; *inside* callipers, for inside measurements, have two hinged and tapered legs finishing with points which turn outwards.

GAUGES are used to mark one or more lines on pieces of timber which are parallel to an edge ; the varieties include the marking gauge, cutting gauge, mortise gauge and panel gauge.

Marking Gauge (see 4) has a holed beech head which receives a box-wood stem, near one end of which is a sharp steel marking point (called the *pin, spur* or *tooth*) which projects 5 mm below the lower surface ; the stem can be fixed in any desired position by means of the thumbscrew. After the stem has been set as required and the screw tightened, the face of the head (that nearest the point) is placed against the edge of the timber, the point is pressed down firmly and this scores a line on the surface as the head traverses the edge.

Cutting Gauge.—This is similar to the marking gauge, except that it has a steel cutter in place of the spur ; the cutter is about 35 mm long, 5 or 6 mm wide and 2 mm thick, sharpened to a point ; used for cutting parallel strips from thin stuff, such as veneers, and for marking across the grain.

Mortise Gauge (see 6).—This has two marking points, that nearest the end of the stem being fixed and the other movable ; the latter is attached to one end of a brass slide which is dovetailed into the stem and extends from the fixed pin to a collar which is fitted over the brass screw at the opposite end ; this screw penetrates the stem, and as it rotates it withdraws the slide (and the movable pin) to increase the distance between the two pins ; this distance may be adjusted from 20 to 50 mm ; the head or stock is fixed in the desired position to the stem by the brass thumbscrew shown at the top of the head. The gauge thus enables two parallel lines to be marked, and is employed for setting out mortises and tenons (such as are required for doors) when, for example, the points are set to the width of the mortise and the head is then adjusted to the required distance from the movable pin.

Panel Gauge.—This is larger than but resembles the marking gauge ; usually made by the joiner, the stem being 20 mm square by about 700 mm long, with a 60 mm wide by 25 mm thick by 200 mm long head. The pin is fixed, and the head is adjusted and screwed tightly as described for the marking gauge ; used in the construction of door panels.

2. CUTTING AND PLANING OR SHAVING TOOLS (see Fig. 156).—These include saws, chisels, gouges, planes and spokeshaves.

SAWS.—There are many varieties, including the cross-cut saw, rip saw, tenon saw, dovetail saw, compass saw, pad saw and bow saw. A saw has a spring steel blade with a wood (usually beech or apple wood) handle securely riveted to it ; the lower edge or front of the blade is divided into serrations or fine teeth ; this cutting edge is usually specified according to the number of *points* (not teeth) per 25 mm ; thus, at A, C and D, the number of points per 25 mm is six, four and ten respectively. The teeth are bent alternately to the right and left of the blade to enable it to pass through the cut being formed in the timber with the minimum of friction as the sawing proceeds ; this bending of the teeth is called the *setting* (see B) and causes the teeth to form a slightly wider cut in the timber than the thickness of the blade. In addition, the blades of the larger saws are ground thinner at the *back* (opposite edge to the teeth) than at the cutting edge. A saw should be as thin as possible, otherwise an excessive waste of material would result because of the wider cut which would be formed by the thicker blade.

Cross-cut or Hand Saw (see I).—This is essentially used for cutting across the fibres of the wood, but it is also used with the grain, and in carpentry for general sawing. It is made in sizes of 55 to 70 cm (length of blade), advancing by 5 cm, a 66 cm blade being standard ; the number of points is 5, 5½, 6, 7 and 8 per 25 mm ; the eight-point saw is considered best for hardwoods, a seven-point saw for both hardwoods and soft-woods and a five-point saw for rough carpentry. The teeth are shaped as shown in the enlarged sketch at A. The hollow back improves the appearance.

Rip Saw.—This is rarely used, unless machinery is not available, when it is employed for cutting timber along the grain ; it resembles the cross-cut saw, is 70 cm long and has teeth shaped as shown at C, with four points per 25 mm.

Panel Saw.—Is similar to the cross-cut saw, but the blade is finer, and the teeth are usually shaped as shown at A ; a 66 cm blade with ten or twelve points per 25 mm is considered best ; is used in preference to the cross-cut saw for accurate work, and instead of the tenon saw (see below) for cutting panels and similar wide work.

Tenon Saw (see 7).—Used for finer work than both the cross-cut and panel saws ; as implied, is used for the cutting of shoulders (see 2, Fig. 157) to tenons and when a clean cut is necessary ; made in 30, 35 and 40 cm sizes, the 35 cm blade being generally preferred ; its very fine blade is stiffened and increased in weight by the brass or steel back or bar, which is tightly pinched on the upper edge. It has ten or twelve points per 25 mm, and the teeth (called *peg teeth*) are of equilateral triangular shape as shown at D ; is sometimes called a *back saw*.

Dovetail Saw (see 2).—This has an 20, 25 or 30 cm blade and an open handle ; like the tenon saw, it has a brass or steel back ; used for very fine work, as for forming dovetail joints in drawers and cutting shoulders on narrow rails ; this fine blade has fourteen points per 25 mm.

Compass or Turning Saw (see 3 and 4).—It is used for cutting curves. One type has the blade riveted to the handle, and another has one or more slotted blades (as shown) which are fitted and screwed to the handle ; the latter type is usually provided with three blades of different sizes which are interchangeable. The teeth are shaped as shown at C.

Pad or Keyhole Saw (see 5).—Is useful for forming keyholes and similar curved work ; it is the smallest saw, the blade tapering in width from 10 mm to 6 mm; the pad or handle is slotted so that the blade can be passed through when not in use; when required, the blade is drawn out to the desired amount (see the broken lines) and tightened by two screws; the teeth are similar to those of the compass saw.

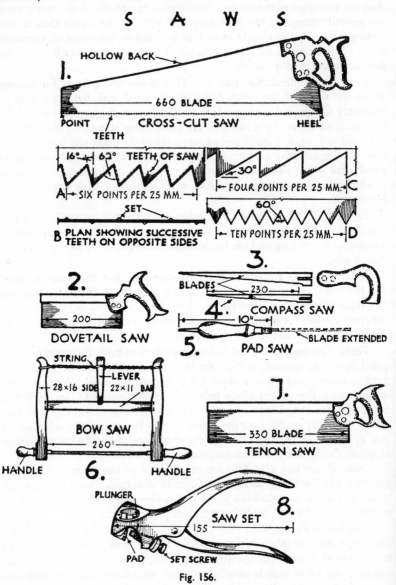

SAWS

1. HOLLOW BACK — 660 BLADE — POINT — TEETH — CROSS-CUT SAW — HEEL

A — 16° — 62° TEETH OF SAW — SIX POINTS PER 25 MM.
C — 30° — FOUR POINTS PER 25 MM.
B PLAN SHOWING SUCCESSIVE TEETH ON OPPOSITE SIDES — SET
D — 60° — TEN POINTS PER 25 MM.

2. DOVETAIL SAW — 200
3. BLADES — 230 — COMPASS SAW
4. 10" — PAD SAW — BLADE EXTENDED
5. BOW SAW — 260 — STRING — LEVER — 28×16 SIDE — 22×11 BAR — HANDLE — HANDLE
6.
7. TENON SAW — 330 BLADE
8. SAW SET — PLUNGER — 155 — PAD — SET SCREW

Fig. 156.

Bow Saw (see 6).—Used for cutting curved work with sweeps which are too quick to be negotiated by the compass saw; the frame is made by the joiner, and the blade (3 to 13 mm wide and 200 to 350 mm long) is fitted into the shafts of two boxwood handles ; about 3 m of strong string is wound round the upper ends of the sides (of ash, beech, birch or mahogany) and a short wood lever is passed between it ; the blade is tightened to the required tension by twisting the string by means of the lever ; this shortens the string, and the blade is stretched as the upper ends of the sides are brought together ; the bar acts as a fulcrum and restrains the lever to prevent the unwinding of the string.

Frame Saw.—Is similar but longer and stronger than the bow saw.

SETTING AND SHARPENING SAWS.—The setting or bending back of the teeth has been referred to on p. 205. The points become worn as the saw is used and the amount of set is reduced ; this decreases the width of the saw-cuts, and freedom of the blade is restricted as the clearance becomes less ; the teeth have then to be re-set and sharpened. Re-setting is carried out before the teeth are sharpened, although it is only necessary to set once to every third or fourth sharpening operation.

The teeth should be of regular length ; if they are not, any correction is made before the setting is commenced ; any irregularity can be seen by looking down the teeth from the *heel* (see 1) and removed by drawing a flat file two or three times along them ; this is known as *breasting*.

The teeth are then set. The expert does this by holding the blade on a steel block or *set* (which has a bevelled edge) and uniformly tapping alternate teeth in turn with a special thin hammer. Alternatively, a patent *saw set* (see 8) may be used ; the saw is gripped between the pad and set screw, both of which are regulated to produce the required set ; the handles are squeezed together, and this causes the plunger to press forward to bend each alternate tooth in turn ; the saw is reversed and the teeth on the opposite side are set in a similar manner.

The saw is now ready for sharpening. It is fixed between the jaws of a *saw vice* (or *sharpening block* or *saw chops*) with the edge of the saw horizontal and about 25 mm above the jaws. A file (triangular in cross section and of a size depending upon that of the teeth) is used ; starting at one end, alternate teeth (those pointing away from the filer) are filed with the same number of strokes of equal pressure and in a manner which varies with the type of saw ; the saw is reversed in the vice and the remaining teeth are filed similarly to bring them to a fine point. Bright points indicate blunt points, and sharp points are dull and difficult to see ; if therefore the teeth have been breasted, each point will be bright owing to its blunt condition, and the filing should be continued until this brightness has disappeared.

CHISELS (see Fig. 157) are tools with forged steel blades in wood (ash, boxwood or beech) handles ; each blade is ground on the back to form a fine cutting edge. They are used to remove thin layers or shavings of wood in shaping surfaces, forming mortises, grooves, etc. The finer the edge, the smoother the surface, and it is essential that the cutting edge be kept sharp by rubbing the back down on an oilstone. Various kinds include the paring, firmer and mortise chisels.

Paring Chisel (see 1).—The *tang* (pointed end) of the thin blade is fitted into the handle, and the brass *ferrule* prevents the tang from splitting the handle ; is used for paring (shaving) plane surfaces both in the direction of the grain and on the end grain of the wood ; the blade may have either square or bevelled edges, the latter type (shown in the figure) being useful in forming rooves ; obtainable in lengths varying from 23 to 53 cm and in widths of from 6 to 50 mm.

Firmer Chisel (see 2).—Is a stronger or " firmer " chisel than the last mentioned, as it has to withstand the action of the mallet (see 4, Fig. 160) which is used to drive forward the tool ; is useful for general work and removing wood in thin chips ; the length varies from 100 mm upwards and the width from 2 to 50 mm.

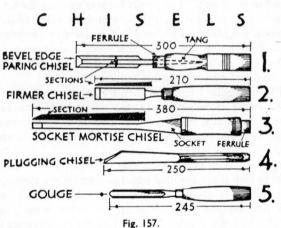

Fig. 157.

Mortise Chisel.— Used for forming mortises, and is much stronger than the firmer chisel, as it has to resist the action of the mallet and the strain resulting in loosening the wood core when making the mortise ; that shown at 3 is known as a *socket mortise chisel* because the wood handle is fitted into the socket of the cast steel blade ; the metal ferrule protects the handle from damage by the mallet ; the ordinary mortise chisels are 3 to 20 mm wide and the maximum width of the socket type is 40 mm.

Plugging Chisel (see 4).—Made entirely of forged steel and used for preparing holes in brickwork, etc. for wood plugs (see pp. 96 and 141).

Pocket Chisel.—Is a very fine chisel, sharpened at both sides, which is used for forming *pockets* in pulley styles of boxed window frames for access to the weights (described in " Joinery ") ; obtainable in widths varying from 40 to 65 mm.

GOUGES are curved chisels which produce circular cuts. Paring, firmer, socket, etc. gouges are available ; that shown at 5, Fig. 157, is known as an *outside ground gouge* and is used for heavy work ; those ground on the inside are used for paring and scribing ; widths vary from 3 to 40 mm.

PLANES (see Fig. 158) are so called as they are chiefly used for shaving or smoothing plane surfaces after the timber has been sawn. They are of (a) wood (beech) and (b) metal (cast steel, gunmetal and malleable iron).

(a) WOOD PLANES.—There are many kinds, including the jack plane, trying plane and smoothing plane (all known as *bench planes*); these are essential items of a kit. Some of the other planes are not so important and may only be used on rare occasions.

Jack Plane (see 1).—This is the first plane used on a piece of timber after it has left the saw; it removes the saw marks and leaves the surface sufficiently smooth for the subsequent finishing with the trying and smoothing planes (see p. 87); it is also useful for quickly planing off large quantities of wood to reduce the scantlings. This plane consists of a stock, double irons, wedge and handle.

The standard beechwood *stock* (or body) is 430 mm long by 75 mm by 75 mm. It should be carefully selected with the annual rings parallel with the face or *sole* (see sketch), otherwise unequal shrinkage will take place, the face will wear unevenly and so affect its accuracy; a reduction in wear results if the grain of the wood runs from *nose* (front) to *heel* (back). The handle is glued into a slot, and a hole is formed to receive the irons and wedge. The width of the *mouth* (see J) is about 50 mm, and a space is left between the irons and the front of the mouth to allow the shavings to escape at the *throat*. A 20 mm hardwood *stud* (or *button*) is fitted on top near to the nose of the plane, and prevents disfigurement as it receives the blows from the hammer when the irons are being adjusted.

The *irons* consist of a *cutting iron* (E) and a *back* or *cap iron* (F) which are made of crucible cast steel. They are made in 50, 54, 58, 62 and 65 mm widths, the 58 mm size being popular. The bottom edge of the cutting iron is rounded, as it is required to remove shavings which should be thickest in the centre and finer at the edge; this edge is double bevelled (see enlarged section through the edge at G), the *grinding bevel* being slightly hollow ground and approximately 25°, whilst the *sharpening* angle is about 32°; the thickness of the iron increases from 2 mm at the top to about 4 mm at the top of the grinding bevel; the iron is slotted to allow movement of the screw which attaches it to the back iron. The back iron (F) of uniform thickness of about 3 mm down to about 13 mm from the bottom, when it is slightly curved back and reduced in thickness to a fine edge; a brass nut is attached to the iron and receives the screw which connects both plates together (see J). The distance that the edge of the cutting iron projects beyond that of the back iron is called the *set* of the iron, and this depends upon the character of the wood to be planed and the thickness of the desired shaving; the set is approximately 3 mm for softwoods and 2 mm for hardwoods. The object of the back iron is to deflect the shaving and bend it as it proceeds through the mouth.

The irons are secured in the stock by means of a wood *wedge* (see 1, H and J). The wedge is knocked down by a hammer when fitting the irons, and, as it passes down the back iron, the fine tapered legs proceed

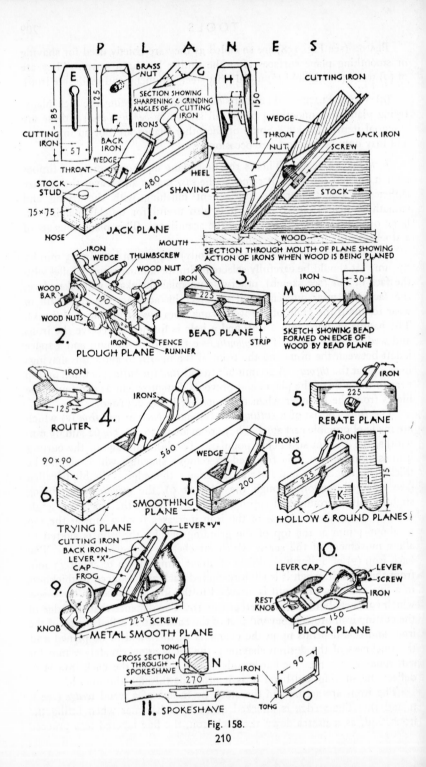

P L A N E S

E CUTTING IRON 185
125
CUTTING IRON 51
THROAT

BRASS NUT

G SECTION SHOWING SHARPENING & GRINDING ANGLES OF CUTTING IRON

H 150

CUTTING IRON

WEDGE

F IRONS
BACK IRON
WEDGE

BACK IRON
THROAT NUT
SCREW

STOCK STUD
HEEL
480
SHAVING

J STOCK

75 × 75

NOSE

1. JACK PLANE

MOUTH
WOOD

SECTION THROUGH MOUTH OF PLANE SHOWING ACTION OF IRONS WHEN WOOD IS BEING PLANED

IRON
WEDGE
THUMBSCREW
WOOD NUT
WOOD BAR
190
WOOD NUTS
IRON FENCE
RUNNER

2. PLOUGH PLANE

IRON
225

3. BEAD PLANE
STRIP

M IRON 30
WOOD

SKETCH SHOWING BEAD FORMED ON EDGE OF WOOD BY BEAD PLANE

IRON
125

ROUTER **4.**

IRONS
560
90 × 90

6. TRYING PLANE

IRONS
WEDGE
200

7. SMOOTHING PLANE

5. IRON 225
REBATE PLANE

8. IRON 225
K
L 75
HOLLOW & ROUND PLANES

LEVER "Y"
CUTTING IRON
BACK IRON
LEVER "X"
CAP
FROG
9.
KNOB
225 SCREW
METAL SMOOTH PLANE

10.
LEVER CAP
LEVER
SCREW
IRON
REST KNOB
150
BLOCK PLANE

TONG
CROSS SECTION THROUGH SPOKESHAVE
N
IRON
270
90
O
TONG
11. SPOKESHAVE

Fig. 158.

down the two side grooves in the hole in the stock until the irons are rigidly fixed.

Trying Plane (see 6).—This is used for precise work, such as removing irregularities left on the surface of the timber by the jack plane ; it is also used for forming long straight edges, as for joints and nosings, and traversing wide widths. It is the largest bench plane (the sizes being 56, 60 and 66 cm and closely resembles the jack plane but for the closed handle. The set of the irons is usually 2 mm for softwoods and 1 mm for hardwoods.

Smoothing Plane (see 7).—This is known as the *finishing plane*, as it is used to smooth the surface of the timber after the jack and trying planes have been applied. The stock is only 20 cm long and is provided with double irons set as for the trying plane.

Rebate Plane (see 5).—Used for forming rebates, and has only a single iron. This cutting iron is fixed by a wedge and is the full width of the stock ; the iron, which varies from 6 to 50 mm in width, is placed either on the skew (" skew mouth ") or square (" square mouth "), the former being preferred as it is considered to work easier.

Hollow and Round Planes (see 8).—The hollow plane is used for producing convex surfaces on the timber (see enlarged section through the sole at κ) and the edge of the single iron or cutter conforms to the curve. Concave surfaces on timber are formed by the round plane (see enlarged section L).

Formerly, a joiner had as part of his kit at least a half set of " hollows and rounds," and many possessed complete sets consisting of eighteen pairs, varying from 3 to 40 mm in width, for the purpose of making mouldings. Nowadays most of this work is done by machinery, and these hand tools are but rarely used.

Bead Plane (see 3).—This moulding plane is still required, and two or three different sizes should form part of a kit ; it is used for forming a half-round moulding with a *quirk* (sinking) on edges of members. The strip let into the sole of the stock is of boxwood to resist wear. A sketch showing the application is given at M.

> *Note.*—A number of moulding planes, such as ogee, torus, reed, astragal, ovolo, etc. have practically fallen into disuse, since mouldings can be produced much more cheaply by machinery.

Plough Plane (see 2).—This is used for forming grooves with the grain, varying in width from 3 to 16 mm, and to any depth up to 30 mm ; the single iron, secured by a wedge, passes down a narrow mouth formed in the metal *runner* or guide screwed to the stock ; the depth of the groove is regulated by the metal thumbscrew which passes through the stock and depresses or raises a metal solepiece (about 20 mm wide) which operates between the runner and wood *fence* ; the wood nuts which negotiate the wood screw bars are manipulated to adjust the width between the fence and the runner as required. The plough is provided with six or eight irons of different widths.

Router or Old Woman's Tooth (see 4).—This plane is used for increasing to a uniform depth grooves (an operation known as trenching) formed previously by another tool ; the strong iron is from 3 to 13 mm wide.

Spokeshave (see 11).—Used for planing circular work having quick curves ; the iron (see O), which should be well sharpened, is fixed by passing the two tapered tongs or tangs through the stock ; it is adjusted by lightly tapping either the projecting ends of the tongs or the blade as required (see section at N).

Compass Plane.—This is a smoothing plane with a curved sole and 50 mm wide double irons for planing curved surfaces. It is not much used.

Moving and Sash Fillister Planes.—Both are used for forming rebates or grooves. The moving or *side fillister* is a rebate plane with a movable fence, a single cutting iron, and a small side iron for marking out the rebate ; is adopted for making rebates on the near side of the stuff. The sash fillister resembles the plough, and may be used for forming sinkings on the back edge of the wood. These planes are now seldom used.

Toothing Plane.—Is a useful tool for preparing surfaces of timber which are to be glued together, such as is required for veneering ; its 50 mm wide single iron has a serrated edge. The surface to be toothed is levelled by a jack plane and traversed with the toothing plane (diagonally in opposite directions and finally with the grain) to give a flat rough surface.

Tonguing and Grooving Planes (also known as *matching planes*).—Used to form tongues and grooves on the edges of boards required for matchboarding (for doors), battened doors, etc. Although most of such work is now done by machinery, these planes are occasionally required, especially when preparing work during fixing.

(*b*) METAL PLANES.—Most of the aforementioned wood planes are also obtainable in metal, such as cast steel, gunmetal, malleable iron or aluminium. Some of them are an improvement upon the wood planes, but the wood jack plane especially is still considered the best for the purpose. The metal planes are more fragile than those in wood, and therefore the jack and trying planes are better able to withstand the somewhat rough usage to which they are often subjected. Metal planes are also more expensive than wood planes. Very accurate work, especially in hardwood, can be produced by metal planes, and their various adjustments can be readily effected. The sole of a metal plane is not subjected to the wearing action common to the wood plane.

Metal Smooth Plane (see 9, Fig. 158).—This is a very useful tool, especially for smoothing the surfaces of hardwoods of best quality which have been previously dressed with the jack and trying planes. The *cap* secures the two irons (called the *cutter*) by a screw which passes through to the *frog* that supports them ; the cap is adjusted by the lever " x " ; the lever " Y " adjusts the cutter sideways, the frog is regulated either forward or backward by an adjusting screw, and the large screw or *milled nut* behind the frog adjusts the edge of the cutter to regulate the thickness of shaving. This tool is the Stanley " Bed Rock " smoothing plane, and is obtainable in sizes varying from 140 to 250 mm in length of sole with cutters which are from 30 to 60 mm wide. A somewhat similar tool, called the *English steel smoothing plane* and having a width of iron up to 60 mm, is very satisfactory, although it is not so readily adjusted as that described on the previous page.

Block Plane (see 10).—This is a very desirable tool, especially for small work which is not readily accessible, and for preparing mitres of hard-wood mouldings. It is well suited for planing against the grain, and, on account of its small size, it can be easily gripped and controlled by one hand. It has only a single iron or cutter which is inclined at 12° to 20° (compared with the 45° pitch of the cutting iron of the wood jack, etc. planes—see J, Fig. 158) ; unlike the cutting iron of the wood bench plane, the bevel of the cutter of the block plane is uppermost. The type shown is a Stanley plane, and to assemble it, the iron (which has a central slot) is placed over the small projecting lever cap screw ; the cap (which has a knuckle joint) is fitted over it, and when correctly placed, pressure on the cap springs it into position ; the edge of the cutter is brought parallel with the mouth (barely 6 mm wide) by lateral movement of the lever, and the distance between the edge of the cutter and the front of the mouth is regulated as required by the milled screw or nut shown below the lever ; the width of the iron is 40 mm.

Other varieties of metal planes include the *bullnose plane* (the edges of the iron is close up to the nose of the plane, and is therefore useful for planing surfaces at the ends of rebates, etc.) and the *shoulder plane* (which is a form of rebate plane used for planing rebates in hardwood and particularly the ends of members such as the shoulders of door rails).

3. BORING TOOLS (see Fig. 159).—These include the brace and bits, auger, gimlet and bradawl.

Brace and Bits (see 45 to 50).—A brace is a handle or stock to which is attached a cutter or *bit* used for boring holes. Hand pressure on the *head* of the brace assists the boring action of the bit whilst the brace (gripped by the handle) is revolved. That shown at 45 is of the *ratchet* type and is the best, for, when desired, the turning movement of the handle may be restricted to a small arc to allow boring in confined positions. When the ratchet is suitably adjusted, the bit only bores into the wood when the brace is turned clockwise through a part of a circle, and the bit remains stationary when the brace is turned backwards. The brace with the screwdriver bit attached is also employed to force in screws when pressure on the ordinary screwdriver (see 5, Fig. 160) would be inadequate. The *chuck* contains steel spring jaws into which the shank of the bit is inserted and securely gripped by rotation of the chuck. The *sweep* of the brace is of steel, the head and handle are of hardwood, and in the best tool the head, handle and ratchet head have ball bearings providing easy action.

There are many varieties and sizes of bits. The *centre bit* (46) is employed for boring ; the cutting edge P cuts out the circumference of the hole as the bit is rotated by the brace, and the turned back cutting edge Q removes the waste material from the hole; its diameter varies from 3 to 40 mm. The *shell bit* (50) resembles the gouge (compare with 5, Fig. 157), the *spoon bit* (which resembles the gouge, but pointed), the *nose bit* (like the shell bit, but with a cutting edge at the point) and the *screw bit* or *twist bit* (which has a screw thread at its point) are used for producing small holes from 3 to 13 mm diameter. *Auger* or *twist bits* form holes which are cleaner and more accurate than those produced by the above varieties; there are many patterns, such as Russell Jenning's (48), Gedge's and Irwin's (49), etc.; these are in two lengths, the shortest being known as *dowel bits*, and the diameters increase by 2 mm from 6 to 40 mm. The *Forstner bit* is unlike the twisted bits, as the end has a circular rim instead of a point; the larger bits have only plain and not spiral shanks; it is useful for boring in any direction. The *expansion bit* is provided with adjustable cutters of different sizes; thus, in one range the cutter can be adjusted to bore holes varying from 13 to 40 mm diameter. The *screw-*

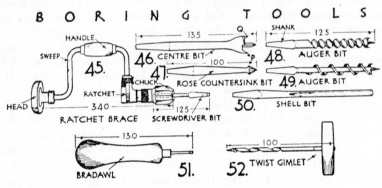

Fig. 159.

driver bit (45) is an important tool and has already been referred to (p. 213). *Countersink bits* are used to prepare shallow sinkings to receive heads of countersunk screws (such as K, Fig. 152), etc.; the *rose* countersink bit (47, Fig. 159) is suitable for both hardwoods and metals, the *snail* countersink bit (similar to the rose, but having a sharper point and a grooved end) is used for wood only, and the *flat* countersink bit (having a flat end which is tapered to a point) is only suitable for boring through metal. *Rimers* are tapered bits which are used for either preparing tapered or conical-shaped holes or for increasing the size of the holes.

    Auger.—This has a steel stem, about 60 cm long (although this may be exceeded), having a round eye at one end to receive a wood cross handle; the other end is shaped like the bits of this name (48 and 49, Fig. 159): is used for deep borings up to 50 mm diameter.

    Gimlet.—This small tool is useful for boring holes to mark the position and facilitate the insertion of screws. The various patterns include the *twist gimlet* (52, Fig. 159), *shell gimlet* (resembling a gouge with a screw end) and the *auger gimlet* which has an augered shank.

Bradawl or Sprig Bit (see 51, Fig. 159).—The small steel blade has a sharpened end. It is used for making small holes for screws, etc.

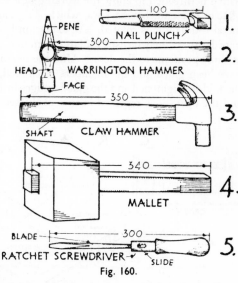

IMPELLING TOOLS

Fig. 160.

4. IMPELLING TOOLS (see Fig. 160) comprise hammers, mallets, screwdrivers and nail punches.

Hammers.—That shown at 2 is called the *Warrington hammer*; the head (usually of cast steel with a tempered steel *face* and *pene*) is wedged to the shaped ash or hickory shaft; of the many sizes, that with the head weighing approximately 0·45 kg is most used for general purposes. The *claw hammer* (3) is made with heads weighing from 0·2 to 0·8 kg; the claw is useful for levering back or withdrawing nails.

Mallet (see 4).—Used for driving chisels and knocking framing together; the tapered mortise in the beech head receives the slightly tapered ash or beech shaft.

Screwdrivers or Turnscrews.—There are three forms, *i.e.*, the *fixed-blade* type, the *ratchet* driver (5) and the *spiral* driver. The former is obtainable with the length of blade varying from 75 to 300 mm and is the firmer tool to employ for heavy framing. The ratchet screwdriver, by adjusting the slide, can be turned right or left without releasing the hand pressure; it can also be converted to the rigid type. The spiral turnscrew has a spiral shank down which the handle is pushed; this drives the screw in or out according to the slide adjustment; a spring in the handle causes it to return quickly for the next push.

Nail Punches (see 1).—These vary in size and shape, and are used to punch the heads of nails below the surface of the timber.

5. CRAMPING AND HOLDING APPLIANCES (see Fig. 161) include T-cramps, G-cramps, bench holdfasts and mitre blocks.

T-cramp.—This is used to cramp up (compressing the joints) framings, such as doors, window frames, etc., during the gluing and wedging process. A sketch showing the application to a door is shown at 2, Fig. 161. The cramp is a steel bar of T-section which is from 45 to 70 mm deep, 20 to 25 mm at its flange or widest part, and from 60 to 210 cm long; it has a series of 13 mm diameter holes along its length into which a 75 mm by 13 mm round steel taper peg is inserted; this peg is attached by a chain to a shoe, the jaws of which pass over the flange of the bar to enable the shoe to slide along it; at the other end of the bar there is a metal head which is threaded to allow the working of a screw which has a rectangular plate at one end, having jaws, which slide along the bar flange when the metal rod handle is rotated. An extension bar may be fitted to the cramp in order that it may be used for large framings. The cramp is shown applied at three positions to the door. After the tenons and insides of the mortises have been glued and the members assembled, the cramp is applied along the middle, tightened and the glued wedges inserted; the cramp is removed and the operations repeated in turn at the second and third positions.

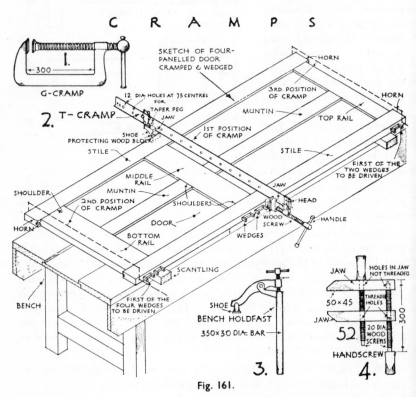

Fig. 161.

G-*cramp* (see 1, Fig. 161).—This metal cramp is convenient for small work; the sizes vary from 13 to 30 cm (distance between the end of the screw and the opposite leg). A lighter but similar cramp, with a thumb-screw instead of the lever handle and having a maximum clearance of 20 cm, is also used.

Bench Holdfast or Clamp (see 3).—Is of wrought iron with a steel screw and malleable arm and shoe; the arm varies from 25 to 35 cm long and from 20 to 30 mm diameter. Its object is to grip the stuff on the joiners' bench during the process of working. The bench top is holed to receive the bar, the work is gripped by the shoe, and the shoe is tightened to cause the bar to cant over and grip the sides of the hole ; this forces the shoe down and makes it rigid whilst the joiner is engaged in doing the necessary labours.

Handscrew (see 4, Fig. 161).—This consists of two wood (hornbeam or beech) or metal screws with two beech jaws. It is one of the best appliances for cramping light stuff during the actual hand operations and after the work has been glued, as the comparatively large jaws do not damage the stuff.

Mitre Block (see 4, Fig. 162).—Used in forming mitres on architrave and panel mouldings, etc. ; it consists of two pieces of wood which have been carefully dressed and glued together ; two 45° marks are accurately set out by using the bevel (3, Fig. 155) and try square (2, Fig. 155), and the two cuts or *kerfs* are carefully formed with a tenon saw (7, Fig. 156) ; sometimes a square cut is also formed for butt joints. In cutting the mitre, the length of moulding is placed on the block with the moulded face outwards, the saw is placed in the cut, and the moulding is sawn with the kerf serving as a guide.

A *mitre box* (which is in the form of a channel and consists of two 25 mm vertical pieces of wood secured to a wood bed piece) is sometimes used for large mouldings ; mitre cuts are made down the two vertical pieces. The large moulding is placed within the box and made rigid by wedges ; the tenon saw is placed across the box and engaged in the two short cuts, and the mitre is sawn down the moulding.

A *mitre templet* (used for trimming the mitres after being cut) and *shooting and jointing boards* (employed in planing mitres and edges with the trying plane—6, Fig. 158) are other forms of equipment.

6. ABRADING TOOLS include scrapers and rasps.

Scraper (see 7, Fig. 162).—The two longer edges of this steel plate (which does not exceed 2 mm in thickness) are turned over to form a slight *burr* (keen-cutting projecting edge) on each side ; it is used on a hardwood surface after the latter has been levelled and smoothed by planing ; the scraper is gripped by both hands and the burred edge drawn or pushed over the surface of the timber, in varying directions (finishing with the grain) until any defects left by the planes have been scraped out.

Rasps.—Two grades of the steel *half-round rasp*, shown at 6, Fig. 162, are used to remove bumps on curved surfaces; the coarse and fine files are about 25 and 20 cm long respectively. The fine file eliminates the marks left by the coarser file. Flat rasps are also obtainable.

Glass-papering, also termed *sand-papering*, is the final process to good hardwood surfaces. Thus, after a surface has been planed by the jack, trying and smoothing planes and scraped, it is traversed (generally with the grain) by the *rubber* (a piece of cork round which is wrapped a piece of glass-paper). This abrading material is a strong paper coated on one side with a mineral like garnet, glue being used as an adhesive ; obtainable in various grades ; usually application of two or three of the grades is necessary before the surface is completed. Mahogany and certain other hardwoods should be " damped down " (*i.e.*, the surface is damped with a little hot water) and allowed to dry before the finer grade of glass-paper is applied ; this is necessary to " raise the grain " which has been depressed by the action of the coarser paper.

MISCELLANEOUS APPLIANCES

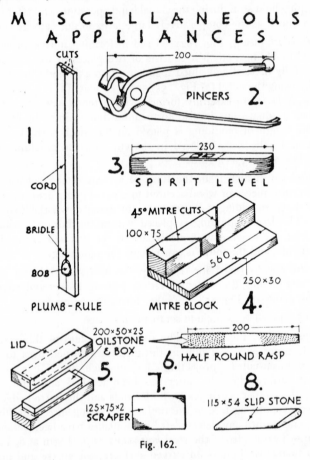

Fig. 162.

Oilstone and Box (see 5, Fig. 162).—There are several natural and artificial oilstones, and these vary considerably in degree of fineness. Well-known varieties are the Arkansas, Carborundum, India, Washita and Turkey. A good quality oil (preferably neatsfoot) should only be used when sharpening the tools. The hardwood (mahogany) box to receive the stone, and the lid should be made out of the solid.

Slip Stone (see 8, Fig. 162).—These are of similar materials to oilstones and are used for sharpening gouges ; sizes vary from 10 to 15 cm long, 25 to 60 mm wide and 10 to 20 mm thick.

TOOLS

Grindstone.—Plane irons, chisels, etc. have to be ground before being finally sharpened on the oilstone. A hard grit stone (sandstone) or carborundum, etc. disc, which is caused to rotate during the grinding operation, is generally provided in the shop for this purpose.

7. MISCELLANEOUS TOOLS AND EQUIPMENT (see Fig. 162).—Other tools necessary in a joiner's tool chest include :—

Cold Chisel.—Is a strong steel tool, about 13 mm wide, which is used for the removal of superfluous plaster, prior to the fixing of architraves, skirtings, etc.

Pincers (see 2).—That shown is the older type, and, as the claw is seldom used, the newer form which is not provided with a claw and knob is sometimes preferred, as it is considered to give a better grip.

Axe.—This is useful for rough carpentry work.

Plumb-rule (see 1).—Is a dressed piece of wood, 140 to 180 cm long, 9 to 10 or 13 cm wide and 13 to 25 mm thick, holed near the bottom to permit slight movement of a pear-shaped lead *plumb-bob* which is suspended by a length of whipcord fixed to the top of the rule. The long edges of the rule must be parallel. A centre or *gauge line* is cut down its face and extends the full length. Three short saw-cuts are made down its top edge, one coinciding with the gauge line and one inclined on each side ; the cord is passed down the middle cut and wound round the side cuts two or three times. The bob weighs from 1·3 to 1·8 kg and the hole is slightly larger than the bob. A copper wire *bridle* or *guard* is bent over the line and fixed just above the hole to prevent excessive swinging of the line and bob. The plumb-rule is essential for testing vertical members of framing (such as door frames) that is being fixed.

Spirit Level (see 3).—Used in conjunction with the straight-edge (7, Fig. 155 and p. 84) for obtaining horizontal surfaces. Consists of a hardwood case, in the central recess of which is fixed a glass tube filled with alcohol (save for a small bubble of air) and having a slightly concave lower surface. A horizontal or level surface is denoted when the spirit level placed on it (or on the middle of the straight-edge when applied on its edge) shows the bubble in the centre.

Oil Can.—The " non-leak " cone-shaped type is preferred.

Tool Box.—The apprentice must provide himself with a tool box, made of wood, in which to keep his tools. The box is approximately 20 cm wide by 30 cm deep by 85 cm long. A 5 cm deep tray should be fitted inside near the top to hold chisels, gouges and small articles ; it is divided into two or three compartments. The lid of the box generally forms the front face and is fitted with bearers to hold the hand saw and panel saw. A lock is fitted and one or two handles are fixed.

A second box, about 15 cm wide by 30 cm deep by 60 or 70 cm long is sometimes provided. This is easier to carry than the above box and will hold sufficient tools for most outside jobs.

Bit Bag or Bit Wrapper.—This is necessary to hold bits and other boring tools, screwdrivers, plane irons, etc. If a tray is not provided in the tool box, chisels and gouges are kept in the bag. It is made of mole-skin cloth or other strong material, size about 60 cm by 45 cm, one long edge is turned up 10 cm to form a pocket which is stitched at intervals, and a piece of wide tape is sewn at intervals lengthwise and about 5 cm above the pocket to form a series of compartments to receive the tools placed with their cutting edges in the pocket. The second long edge is folded over the ends of the tools, the bag is rolled up, and a tape sewn in the middle outside serves to fasten it.

Name Stamp.—This is needed to stamp the name of the owner on his tools to enable him to identify them. It is made of forged steel, with the name in small (3 to 5 mm high) letters. Care should be taken when stamping the tools to avoid damage to the handles.

8. PORTABLE POWER TOOLS (see Fig. 163).—Five [1] of these electrically operated tools are shown at 1 to 4 and 8 in this figure; they are provided with a three-core, tough rubber-sheathed cable for plugging into the power supply. The tool at 6 [2] is powered mainly by a small cartridge explosive.

Electric Saw (see 1).—This saw is mounted on a sole plate which rests on the timber being cut. The blade is 18 cm in diameter, giving a maximum vertical cut of 6 cm; other blades are available, including a planer blade for smoothing timber and a silicon carbide abrasive disc for cutting stone, brick, cast iron, bronze and asbestos; an aluminous oxide disc is used for cutting steel and high tensile materials. A front and rear handle, which incorporates a trigger switch, are provided for the operative to push the saw over the material being cut. The sole plate has an angle adjustment to give bevel cuts and another adjustment to regulate the cutting depth. Under load, the blade revolves at 3,000 revolutions per minute with an input of 1,050 watts. When used for ripping lengths of timber, a guide can be fitted on to the two screws at the side of the sole plate, the guide slides along the edge of the timber and maintains the correct width of cut. A similar model with a 25 cm blade is also obtainable.

Electric Belt Sander (see 2).—One-hundred mm-wide sanding belts of different grades to suit the hardness of the material being smoothed are fitted to this tool. An adjusting knob is provided to centralise the belt and a bag is attached to collect the dust. It has a front and rear handle in which the switch is located. The time spent in smoothing down timber surfaces is greatly reduced by this tool, the belt speed being 335 m per minute on light load, with an input of 775 watts on full load.

Electric Screwdriver (see 3).—In this tool, a pilot hole is bored in the wood first, the screw is then offered up to the hole and the head of the

[1] Manufactured by Wolf Electric Tools Ltd., the electrical mechanism of all these tools is similar to that described under the 6 mm *electric drill* (see below).

[2] Manufactured by MEA-Aktiengesellschaft, Schaan, Liechtenstein, and marketed in this country by Wordrew Ltd.

tool containing the screwdriver bit inside the sleeve guide is placed over the screw. On depressing the switch in the handle the screw is driven home ; a clutch is provided which slips when the required load is achieved so that the screw cannot be driven beyond the desired depth. The model shown has two speeds to give the correct drive for the different materials being drilled. In oak, 10 cm by No. 14 screws can be driven home without difficulty. This tool can also be used for driving hexagon metal thread screws and nuts up to 1 cm. diameter Whitworth. On full load the spindle speed is 290 revolutions per minute with an input of 280 watts.

Electric Rotary Percussion Drill (see 4).—This is used for both normal drilling and percussive drilling, the changeover being made by an adjusting ring on the front of the machine. Four weights of percussion drive are provided to give the correct weight to suit the particular job. Where timber fixings are made to concrete the machine enables the correct action to be given for drilling through the wood, and by adjusting straight into the concrete which requires percussive action for efficient drilling. The drill is double insulated for operator protection—this means that it does not rely on earthing for its safety as the whole body is made of insulating material.

6·4 mm Electric Drill (see 8).—This is a popular general-duty drill and is the forerunner of the other power tools given above. It is capable of drilling 6·4 mm diameter holes in steel and 16 mm diameter holes in hardwood ; on full load the bit rotates at 1,900 revolutions per minute with an input of 280 watts. When the current is switched on it flows through the field coils, setting up a magnetic field causing rotation of the armature which transmits to gears ; these operate a fan for cooling and also a spindle with the attached chuck. The drill is an all-insulated model and, therefore, does not rely upon earthing for safety, the whole body being of insulated material to avoid operator contact with any electrical part. This complete envelope of insulation encloses a special nylon chuck spindle which isolates the metal chuck from the electrical parts. Incorporated in the chuck are three jaws for gripping the bit which can be opened or closed by fitting the key (see 7) into a hole on the outside of the chuck. Serrations on the key engage with those on the chuck, enabling the bit to be secured or released. Various attachments can be fitted to the drill to drive a small circular saw, lathe or grinding and sanding discs ; it can also be fixed to a drill stand or press which is fastened to the bench for drilling vertical holes. Larger similar models of the drill are made for heavier work such as drilling 25 mm diameter holes in steel or 65 mm diameter in wood ; such a model can be readily converted to a mortiser by using a hollow chisel and bit like that at F, Fig. 8, in the companion volume " Joinery."

" *Hilti* " *Cartridge-assisted Tool* (see 6).—This is a percussive tool used for attaching door and window frames, battens, brackets, pipe and conduit clips to brick, concrete and stone. It obviates the drilling of holes in these materials and can also be used for driving fastenings direct into steel. Special hardened steel nails known as *Micky pins* are used, and these are

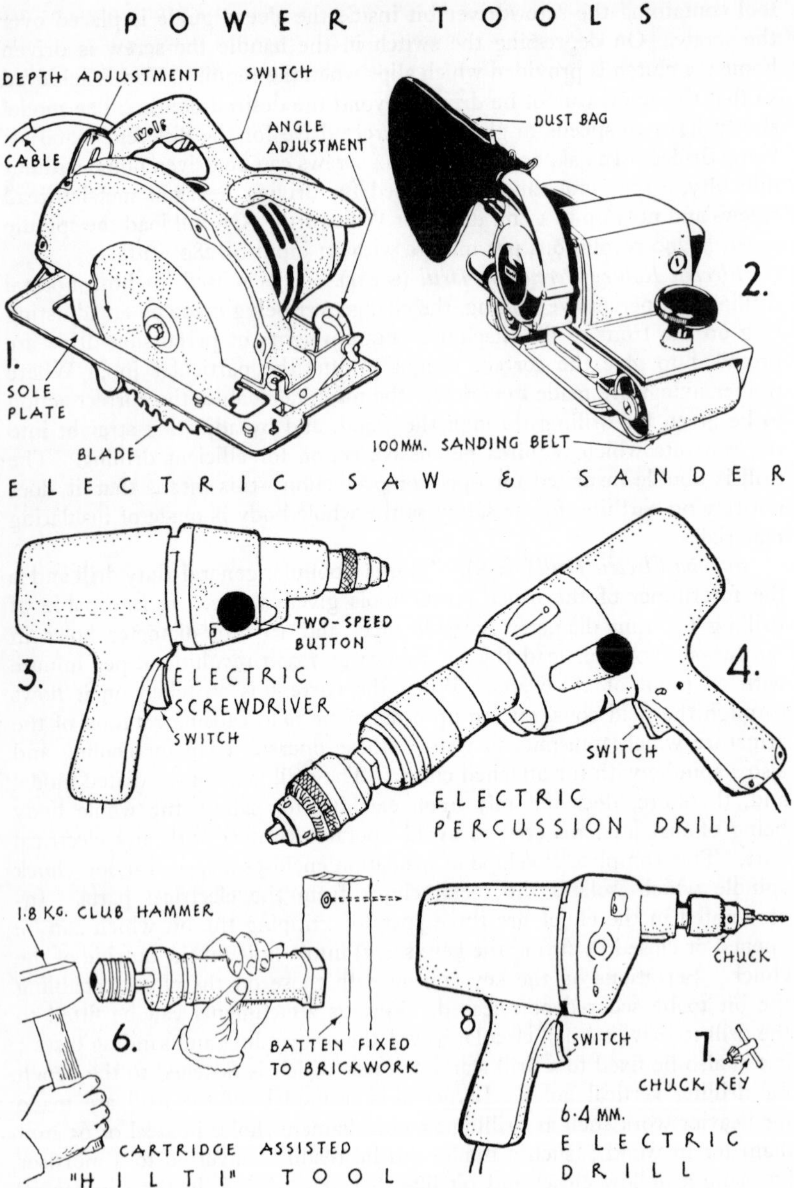

POWER TOOLS

DEPTH ADJUSTMENT   SWITCH

CABLE

ANGLE ADJUSTMENT

DUST BAG

1.

SOLE PLATE

BLADE

2.

100MM. SANDING BELT

ELECTRIC SAW & SANDER

3.

TWO-SPEED BUTTON

ELECTRIC SCREWDRIVER

SWITCH

4.

SWITCH

ELECTRIC PERCUSSION DRILL

1.8 KG. CLUB HAMMER

6.

BATTEN FIXED TO BRICKWORK

CARTRIDGE ASSISTED "HILTI" TOOL

8.

CHUCK

SWITCH

7.

CHUCK KEY

6.4 MM. ELECTRIC DRILL

Fig. 163.

forced through the item being fixed into the brick or concrete backing. Three main types of pin are available:—with rounded head as shown at 6, with threaded end for subsequent fixing of a nut, and with recessed tapped end into which a bolt can be screwed. The tool comprises an outer sleeve containing a barrel wherein a cartridge plunger and a front plunger are enclosed. The outer end of the cartridge plunger has a head for receiving the hammer blow and the internal end is recessed to hold a small cartridge (6 mm diameter). The end of the front plunger is recessed to contain the head of the pin that is being fixed. A blow from 1·8 kg club hammer sends the cartridge plunger forward, making contact with the front plunger. This causes the cartridge to be fired, driving the front plunger forward, thus thrusting the pin into the batten and into the brickwork.

CARE OF TOOLS

Good tools are an essential aid to good workmanship. Only those of best quality should be employed and these should be obtained from a reliable firm. When selecting his tools, the apprentice should be guided by the advice of an experienced craftsman.

After purchase, the tools should be kept in good condition. Ill-conditioned saws, blunt chisels and plane irons, for example, produce badly finished work, and extra effort and longer time are required to accomplish it.

New wood planes, after the irons have been removed, should preferably be soaked in a vat containing raw linseed oil until the wood is well impregnated with the oil ; alternatively, the mouth of each plane should be sealed with putty and the throat filled with oil which is left for about a day. This adds to the efficiency of the planes, prevents the absorption of moisture, and improves the appearance.

At frequent intervals, when the planes are in use, they should be wiped over with a rag that has been soaked in raw linseed oil.

Studs on the jack and trying planes should always be brought into use when releasing the irons to re-sharpen and re-set. Hammer marks on the wood stocks are unsightly.

Soles of wood planes will, in course of time, show signs of wear, and it will be necessary to true up or shoot them. This should always be done with a trying plane in the following manner :—The plane to be shot is placed (with its sole uppermost) into a vice, with the plane irons and wedge in position but the irons knocked back to prevent their edges contacting and damaging the trying plane. The sole should be shot from front to back, and if the stock has been properly selected, the grain of the wood will be found to be running in that direction (see p. 209).

Metal planes should be wrapped in flannel when not in use.

When planes are not in use, the cutting edges should be set back from the face to provide protection for the irons and so prevent snipping. When working at the bench, planes not in use for short periods should rest with the front ends of the soles on a strip of wood fixed across the well of the bench ; this keeps the cutting edges of the irons clear of grit and dirt ; they should never be left on their sides, as the faces of the planes may warp and the cutting edges may accidentally cause bodily injury.

Plane irons should not be allowed to become too dull before resharpening on the oilstone. When sharpening, the whole of the cutting edge should be in contact with the face of the oilstone, which latter should have a good supply of oil on it ; the corners of the cutting edges should be slightly rounded off. Chisels, gouges, etc. should also be kept sharp. The face of the oilstone should be maintained in good condition and not allowed to become glazed.

When not in use, chisels and gouges should be kept in a rack (with their edges downwards), or in trays in the bench drawer or tool box. The cutting edges are thus protected from damage.

When it is necessary to grind plane irons, chisels and gouges, care should be taken not to overheat the cutting edges in the process, or the temper (hardness) of the steel may be reduced. If the grindstone is a natural (grit) stone, a copious supply of water should be applied to the grinding face, otherwise the face becomes glazed or smooth and will quickly overheat the steel. If an artificial (*i.e.* carborundum) wheel, the grinding face should be kept open or " free cutting ", the tool making only slight contact with the wheel. Tools should be kept cool by immersing them in water before they get hot.

Tools should never be allowed to become rusty. Chisels, gouges and all metal tools should be wiped periodically with a piece of flannel dipped in vaseline to prevent rust. A small amount of vaseline can be smeared on metal planes (which may be only used occasionally) to prevent corrosion. Saws should be kept oiled (machine or neatsfoot) and free from any trace of rust.

Saw sharpening is a skilled operation, and is usually done by a saw sharpener. Improperly filed saws will never give good results.

Saws when not in use should have a lath of size to suit the set on the saw, and grooved to receive the saw teeth along the full length of the cutting edge. This is to prevent damage to the teeth while the saw is in the tool box. In the workshop, they should be hung (at the end of the bench). Saw files and rasps should be kept in a separate compartment of the tool box to prevent them damaging the cutting edges of other tools.

The handles of screwdrivers should be maintained in a smooth condition if blisters on the hands of those using them are to be avoided. Care should therefore be taken not to damage the handles by hammering them with a mallet or hammer.

For the care of electrical power tools, many of the above remarks are appropriate. Other more particular comments are as follows :—(1) See that the tool is properly earthed by means of a three-pin plug and have periodic tests of the earth lead ; (2) carry out regular maintenance like lubrication and cleaning in accordance with the manufacturer's instructions ; (3) avoid allowing the tool to become filled with dust ; (4) do not switch it on or off whilst it is under load as this would cause an excessive strain on the windings ; (5) ensure that carbon brushes do not get overworn.

The various machines used in the preparation of timber are described in " Joinery," a companion volume in this Building Craft Series.

CHAPTER ELEVEN

ADHESIVES FOR JOINERY

THE main adhesives [1] (known also as glues or cements) used to stick timber parts together are :—

(a) Synthetic resins

    (1) Phenol/Formaldehyde
    (2) Urea/Formaldehyde (Aminoplastic)
    (3) Resorcinol/Formaldehyde
    (4) Polyvinyl Acetate (PVA)

(b) Casein glues

(c) Vegetable glues
    (1) Soya bean
    (2) Starch derivatives

(d) Blood albumin glues

(e) Animal glues (Scotch glue)

Depending on the location, an adhesive will need to satisfy one or more of the following requirements : (1) completely, or (2) partially resistant to wet ; withstand (3) heat and (4) mycological attack (m.a.) from micro-organisms which can occur in damp places. Hence it is possible to select a glue for any given purpose [2] but the following other factors have to be considered : (i) The thickness of the " glue-line " ; this is dependent on how closely the parts to be joined can be brought together. Accordingly glues are divided into two kinds. *Close-contact* (CC) glues used mainly for plywood manufacture and similar purposes where heavy pressure ensures tight joints and where a glue-line thickness exceeding 0·13 mm can be avoided with certainty ; and *gap-filling* (GF) adhesives used for general joinery and large assembly work when a thin glue-line cannot be guaranteed, they are suitable for glue-lines up to 1·3 mm thick. (ii) The ease of preparation and application of the glue ; for example, an animal glue has to be

---

[1] The following B.S. are available from The British Standards Institution: 745, Animal Glue ; 1203, Synthetic Resin Adhesives for Plywood ; 1204, Synthetic Resin Adhesives ; 1444, Cold Setting Casein Glue.

[2] In B.S. 1204 the following classification is used : WBP—a weather and boil-proof glue which withstands severe weather exposure, heat and m.a. ; it is practically indestructible and more durable than wood. MR—moisture and moderately weather-resistant ; not boil-proof or as durable as WBP ; withstands full weather exposure for a few years ; it resists cold water for a long time and hot water for a limited time ; good against m.a. INT—suitable for interior use ; not boil-proof nor necessarily resistant to m.a. ; it resists damp and cold water for a limited period.

heated before use and other kinds require mixing with a hardener before they are ready. (iii) The storage-life and pot-life (the time for which the glue remains usable after it has been prepared) ; some glues require special storage conditions and must be used within a limited time after mixing. (iv) The length of time for which adequate pressure must be maintained varies with different glues. (v) The resin glues have to be used within prescribed conditions of temperature and humidity. (vi) Some glues stain the timber ; this is obviously undesirable for veneers and other exposed work. (vii) Certain resin adhesives have to be used carefully to avoid the risk of skin infections to operatives.

Knowing the degree of exposure of the joined timbers and whether GF or CC types are needed, further valuable advice can be had from the adhesive manufacturer whose instructions for storage and application must be followed carefully.

The following is a brief description of the different kinds of adhesive listed above :—

(a) (1) *Phenol/Formaldehyde Resin.*—This is a chemically produced resin derived from a reaction between carbolic acid (phenol) and an aldehyde. It is made in WBP, MR and INT grades either GF or CC and is practically indestructible. Both hot- and cold-setting types are made, the former for plywood manufacture and the latter for general joinery. It is not liable to m.a., is fire-resisting and does not normally stain the wood. The resin can be had in liquid or powder form and some kinds require mixing with a separate liquid hardener ; careful temperature control is needed ; the cleaning of tools is difficult it being only soluble in alcohol ; there is a danger of dermatitis to operatives.

A similar adhesive is the melamine/formaldehyde type usually obtained as a white powder resin to which water is added ; it is normally only suitable for CC work and is weatherproof and resistant to m.a.

(a) (2) *Urea/Formaldehyde* (*Aminoplastic*) *Resin.*—This is also synthetically produced and made from a reaction between urea and formaldehyde. It is made in all the grades given in (a) (1) above ; the resin (syrup or powder) and the hardener (liquid or powder) being either combined ready-mixed or available separately for subsequent mixing. The separate application method (both parts being in liquid form) is used for plywood manufacture where the resin is added to one surface and the hardener to the other ; the surfaces are then brought together and if a quick-setting hardener is used the work is completed rapidly. Slower-setting types are used for general joinery. Pot-life varies from 20 minutes to 24 hours for the cold- and hot-setting types respectively. It does not normally stain timber and is good against m.a. and heat. Care must be taken to use the glue at the temperature stipulated by the maker and within the specified time limits. There is less danger from dermatitis than with the phenolic types ; it is water soluble prior to hardening.

(a) (3) *Resorcinol/Formaldehyde Resin.*—This is chemically made from resorcinol and formaldehyde ; it is more expensive than (a) (1) and (2) being less sensitive to temperature, whilst being applied, than these two.

A top class cement classed as WBP ; is normally GF and so useful for general joinery (also for glued laminated work—see Chap. Five) ; proof against m.a. and is water soluble until hardened. The resin is in liquid form requiring the addition of a powder hardener ; it has a long storage-life ; pot-life is from $1\frac{1}{2}$ to 4 hours.

(*a*) (4) *Polyvinyl Acetate (PVA)*.—This is classed as INT ; it is a ready-mixed white syrupy liquid being increasingly used by joiners in lieu of the once universal animal glue. It is non-staining, easy to use but must be stored at a temperature above $4°$ C and not below $16°$ C. Water soluble and immune to m.a. ; it sets within an hour.

(*b*) *Casein Glues*.—Casein is a milk derivative. Rennet, or acids such as hydrochloric, is added to skim milk to hasten the separation and precipitation of the curd. The latter is finely ground after it has been washed, pressed and dried and borax or other chemicals added. Obtained in powder form, to which water is added before use, it is soluble in water. Widely used for general joinery and also for plywood making. It can be described as moderately gap-filling and is satisfactory for glue lines up to 0·8 mm thick. Some caseins are liable to stain certain hardwoods such as oak and mahogany ; some have a limited resistance to water and others none ; they are susceptible to m.a.

(*c*) (1) *Vegetable Glues*.—The main vegetable used is (1) *Soya bean* grown in the U.S. and Manchuria for its oil content. Other oil seed residue glues are obtained from ground nuts. They are made as a white powder (some kinds contain caustic soda and other chemicals) requiring the addition of water for application. They are moderately moisture-resistant having properties similar to casein ; not much used in the U.K. but widely used for the manufacture of Douglas Fir plywood.

(*c*) (2) *Starch Derivative Glues*—This vegetable glue is derived from cassava (tapioca) flour and incorporates some caustic soda. It is applied cold ; has reasonable strength but is only suitable for interior use ; liable to stain some woods ; seldom used in the U.K.

(*d*) *Blood Albumin Glue*.—Produced from blood obtained from slaughterhouses ; some kinds incorporate chemicals like paraformaldehyde. It has fair water-resisting qualities but is subject to m.a. and will stain certain hardwoods. Whilst it is used for the hot-press manufacture of plywood it is seldom encountered in the U.K.

(*e*) *Animal Glues*.—Known also as *Scotch glue* it is the oldest of adhesives but quite unsuitable for external work ; it is non-resistant to heat and m.a. Due to the amount of preparation required before it is ready for use and the fact that it has no gap-filling characteristics it is no longer the most widely used glue although it is very strong. It is prepared from the skins and bones of cattle, horses, etc. The skins are steeped in liquid lime for two or three weeks, washed, dried and the glue (glutin) is extracted by boiling. The bones are cracked in a mill, placed in benzol or other solvent to remove the fat, taken to a steam boiler where the glue is extracted, and finally purified by heating with alum, etc. Another kind is *fish glue* obtained from fish offal.

Animal glue is prepared for use by softening it by several hours' immersion in from two or three parts cold water ; it is then melted by heating in water-jacketed glue pots and applied hot at an approximate temperature of 60° C ; it should not be boiled. It does not stain the wood, although care has to be taken when applying it to sycamore, maple and similar light-coloured woods to prevent discolouration.

INDEX

A

Abrasive disc, 220
Adhesives, 225–8
Arsenical preservatives, 26
Axonometric projection, 60

B

Bearers, gutter, 162
  soffit, 140
Bevels, 142–7
  hip rafter, 144, 146
  jack rafter, 145, 146
  purlin, 146, 147
  spar, 143, 145
Binders, 71, 102–11
Boards, 81–7, 89, 91, 92, 118
  cleaning off, 87
  laying, 84–7
  manufacture, 83
  soffit, 113, 140
Bolts, 107, 109, 110, 129, 152, 169, 200, 201
Building Regulations, 35, 74, 80, 89, 94, 96, 97, 101, 135

C

Cabot's quilt, 172
Carpenter, qualifications, 8
Carpentry, defined, 8
Ceilings, 105, 118
Centering, 180–5
Characteristics, see " Timber "
Cleaning off floors, 87
Colouring drawings, 69
Concrete, 71, 72, 89–91, 96, 97
Connectors, 152, 200
Conversion, see " Timber "
Copper preservatives, 26
  sheets, 112, 113, 115
Corbel brackets, 102, 120
Cramping, 84, 85
Creosote, 26–9, 75

D

Damp proof courses, 79, 80
Defects, see " Timber "
Dogging, 86
Dormer window, 135–7
Draughtsmanship, 56, 57, 61–70
Drawing equipment, 54–6, 57, 61, 62, 68, 69
  compasses, 54
  drawing board, 54

Drawing equipment—*contd.*
  knife, 56, 57
  notebooks, 56
  pads, 56
  paper, 56, 57
  pencil, 55, 57, 68
  pen nibs, 69
  protractor, 55
  rubber, 55
  scales, 54, 61, 62
  set squares, 54
  tee-square, 54
Dry rot, 31–3, 75, 89

E

Eaves, 113, 114, 118, 137–42, 148, 152
  closed, 113, 139–42
  flush, 113, 118, 137, 152
  open, 113, 138, 148
  sprocketed, 114, 139–42, 148

F

Fibrous asphalt felt, 80, 112, 113, 115, 116, 138
Fireplaces, 87–90, 96–99
Firring, 106, 108, 109
Floor boards, 81–7, 89, 91, 92
  cleaning off, 87
  laying, 84–7
  manufacture, 83
  nailing, 86
  joists, 71–6, 79–81, 84, 89–111
Floors, 71–111
  double, 71, 102–6
    boarded, 111
  levelling, 84
  single, 71–102
  soundproofing, 171–3
  trimming, 87–91, 93–9
  triple, 71, 106–11
Folding boards, 86
Formwork, 193–6

G

Girders, 71, 106–9
Glass paper, 87, 217, 220
Glued laminated portal frame, 153–5
Glues, 48, 49, 154, 225–8
Gutters, 120–1, 162–5

H

Hangers, 123, 131, 132, 134
Hardwood margins, 89, 91, 98
Hardwoods, *see* " Timber "
Hip rafters, 114, 128, 130
  bevels, 144, 145
Hyperbola, 159–60

I

Inking in, 68, 69
Insulating boards, 171–6
Isometric projection, 59

J

Jack rafters, 114, 128, 145, 146
  bevels, 145, 146
Joinery, defined, 8
Joints, bevelled haunched, 95, 97
    housed, 96, 97
  birdsmouthed, 120, 126, 128, 137, 140
  bridle, 128, 157
  butt, 82, 110, 141, 199
  cogging, 78, 104, 108, 126
  dovetailed housed, 95, 96
    halved, 123–5, 148
  feathered and grooved, 81, 83, 86
  fishing, 128
  half-lapped, 77
  heading, 82, 83
    splayed, 83
  housed, 77, 141, 166
  lipped mitred, 111
  matched, 82
  mitred, 110, 200
  notching, 78, 106, 120, 148, 170
  oblique tenon, 128
  pinned tenon, 135
  ploughed and tongued, 82, 86
  rebated, 82, 83, 86
    tongued, and grooved, 82
  scarf, 77, 154, 155
  slot tenon, 168
  splayed, 127
  splayed, rebated, tongued, and
    grooved, 82, 86
  square, 82, 110
    housed, 96
  stub tenon, 166, 167
  tongued and grooved, 81, 83, 86
  tongued, grooved, and vee, 140, 170
    mitred, 110, 111
  tusk tenon, 94, 95, 128, 135
  wedged tenon, 168

K

King post roof truss, 115, 156, 157

L

Laminated wood beam, 110, 111, 153
Lead, 112, 113, 115, 117, 118, 162–5
  cesspool, 119, 164
  drips, 118, 162, 164
  gutters, 118, 162–5
  rolls, 118, 162, 163
Lettering, 63–6

M

Micky pin, 220
Moisture content, 21–6, 32, 34, 196

N

Nails, 197–9
Nuts, 200, 201

O

Oblique projection, 59
Orthographic projection, 59

P

Pantiles, 112, 113, 115
Parabola, 156, 159, 160
Partitions, 166–70, 174–6
  soundproofing, 174–6
Perspective, 60
Plugging, 96, 98, 105, 120, 221
Plywood, 44–53, 111
Portal frame, 153–5
Preservation, *see* " Timber "
Prints, 69
Purlins, 114, 125–8, 146, 149
  bevels, 146

Q

Queen post roof truss, 115

R

Ridge, 114, 122, 126, 132, 147
Roofs, 112–65
  covering, 112, 113, 154
  double, 115, 125–35, 147, 148
  erection, 147–8
  hyperbolic paraboloid, 156, 158–61
  runners, 123, 131–4
  single, 115–25
    close couple, 122, 123–31
    collar, 123–5
    couple, 122
    flat, 116–19
    lean-to, 120
      double, 120
    shell, 156, 158–61
  trimming, 134, 135, 137

triple, 115, 149–56
  king post, 115, 156, 157
  queen post, 115
trussed rafter, 149
trusses, 115, 150–3
  built-up, 114, 150–3
  laminated wood, 153–5

S

Screws, 199, 220, 221
Scribing, 141
Seasoning timber, 21–6
Shingles, 112, 113, 115
Shoring, 185–93
  horizontal, 189–91
  raking, 185–9
  vertical, 191–3
Shuttering, 193–6
Site concrete, 80, 81
Sketching, 68
Slag wool, 171
Slates, 112, 113, 115, 137–40
Softwoods, *see* " Timber "
Soundproofing, 171–6
  doors, 176
  floors, 171–3
  partitions, 174–6
Spars, 112, 115, 120–6, 128, 132–42,
  143–9
  bevels, 142, 143
Strutting, 98–101
  herring bone, 99–100, 118
  solid, 100, 101

T

Thatch, 112, 113, 115
Tiles, 89, 112, 113, 115, 139
Timber, characteristics, 35–43
  figure, 36, 37
  grain, 35, 36
  hardwoods, 40–3
  softwoods, 38–40
  texture, 36
  classification, 11
  connectors, 152, 201
  conversion, 16–19, 46
  flat, plain, or slab, 17
  rift or quarter, 17
  rotary, 18, 46
  tangential, 17
  defects, 18, 19, 23, 29–35, 75, 89
  brashness, 30
  brittleness, 30
  checking, 30, 31
    end, 31
    honeycomb, 31
    surface, 31
  chipped grain, 33
  coarse grain, 29
  collapse, 31
  common furniture beetle, 35
  cup shakes, 30

Timber—*contd.*
  deadwood, 29
  death watch beetle, 34
  doatiness, 31
  druxiness, 29
  dry rot, 31–3, 75, 89
  foxiness, 29
  heart shakes, 30
  house longhorn beetle, 35
  knots, 30
  lyctus powder-post beetle, 35
  shrinkage, 18, 19, 33, 34
  splitting, 23, 34
  swelling, 33
  twisted grain, 29
  wane, 34
  warp, 34
  wet rot, 33
  felling, 16
  fire-retarding, 29
  grades, 10, 11
  growth, 15
  hardwoods, 11–43, 74, 81
  ash, 40
  beech, 40, 81
  birch, 40
  black bean, 40
  canary whitewood, 40
  elm, 40, 81
  gurjun, 41, 81
  iroko, 41, 81
  lauan, 41
  mahogany, 41
  makoré, 42
  oak, 42, 74, 81
  obeche, 42
  rock maple, 42, 81
  sapele, 42
  seraya, 43
  sweet chestnut, 43
  sycamore, 43
  teak, 43, 81
  walnut, 43
  kiln, 24, 25
  plywood, 44–53, 111
  battenboards, 53
  blockboards, 53
  box beam, 52, 53
  composite boards, 53
  laminboards, 52
  merits, 51
  metal-faced, 51
  moulded, 51
  uses, 51
  wall boards, 53, 170, 171
  preservation, 26–9, 33, 75
  anti-termite, 26
  non-pressure processes, 27–9
  pressure processes, 26, 27
    empty-cell, 27
    full-cell, 27
  superficial processes, 28
  seasoning, 21–6

Timber seasoning—*contd.*
 artificial, 24
 combined, 24
 natural, 22–3
 softwoods, 11–15, 38, 39, 40, 74, 81,
  188, 195
  Douglas fir, 38, 74, 81, 189, 196
  European larch, 38
  parana pine, 38
  pitch pine, 38, 74, 81, 189, 196
  red pine, 38, 81, 196
  redwood, 74
  Siberian pine, 39
  sitka spruce, 39, 74, 81, 196
  sugar pine, 39
  western hemlock, 39, 81
  red cedar, 39
  white pine, 40
  whitewood, 39, 81, 196
  yellow pine, 40
 structure, 11–15
 uses, 38–43, 74, 81, 189, 196
Tools, 84, 87, 96, 101, 143, 202–24
 auger, 214
 axe, 219
 bevel, 143, 203
 bit bag, 220
 boxes, 219
 brace and bits, 101, 213, 214
 bradawl, 215
 callipers, 203
 care of, 201, 223, 224
 cartridge assisted, 221–2
 chisels, 96, 207–8, 219, 221
  cold, 219
  firmer, 208
  mortise, 208
  paring, 208
  plugging, 96, 208
  pocket, 208
 compasses, 203
 cramps, 215–7
  drill, 221
 electrically operated, 220, 221, 222
 gauges, 204
  cutting, 204
  marking, 204
  mortise, 204
  panel, 204
 gimlet, 214
 gouges, 208, 224
 grindstone, 218, 224
 hammers, 215, 221
 handscrew, 217
 lathe, 221
 mallet, 215
 marking awl, 202
 mitre block, 217
  box, 217
  square, 203
  templet, 217
 mortiser, 221
 nail punch, 215

Tools—*contd.*
 name stamp, 220
 oil can, 219
 oilstone, 218, 224
 planes, 87, 209–13, 221
  bead, 211
  compass, 212
  hollow and round, 211
  jack, 87, 209, 223
  metal, 212, 213, 223
   block, 213
   bullnose, 213
   shoulder, 213
   smooth, 212
  moving and sash fillister, 212
  plough, 211
  rebate, 211
  router, 212
  smoothing, 87, 211, 220
  spokeshave, 212
  tonguing and grooving, 212
  toothing, 212
  trying, 87, 211, 223
 plumb-rule, 219
 power operated, 220–3
 rasps, 217
 rotary percussion drill, 221
 rules, 202
 sander, belt, 220
 saws, 204–7, 220, 221, 224
  bow, 206
  compass, 205
  cross-cut or hand, 205
  dovetail, 205
  electric, 220
  frame, 207
  pad or keyhole, 205
  panel, 205
  rip, 205
  setting and sharpening, 207
  tenon, 205
 scraper, 87, 217
 screwdrivers, 215, 220, 224
 slip stone, 218
 spirit level, 84, 87, 219
 straight-edge, 84, 87, 201
 trammel, 203
 try square, 203
Tracing, 69
Trench timbering, 177–80
Trimming floors, 87–91, 93–9
 roofs, 134, 137
Trussed rafter roof, 149
Truses, *see* " Roofs "

V

Valley rafters, 114
 gutters, 162
Ventilation, floors, 32, 33, 80

W

Wall plates, 75–9, 101, 102, 113, 116, 120, 126, 128, 138, 139, 148
  boards, 53, 170, 171
Walls, cavity, 71, 154
  fender, 89–91
  party, 101, 127
  sleeper, 73, 80
Wood-wool slabs, 154

Woodworking machines, 19–21, 46, 83
  band re-sawing, 21
  circular saw mill, 20
  horizontal log band mill, 19
  planing and matching, 83
  rotary veneer cutter, 46
  vertical log band mill, 20

Z

Zinc sheets, 112, 113, 115